DNA
Microarray Technology and
Data Analysis in Cancer Research

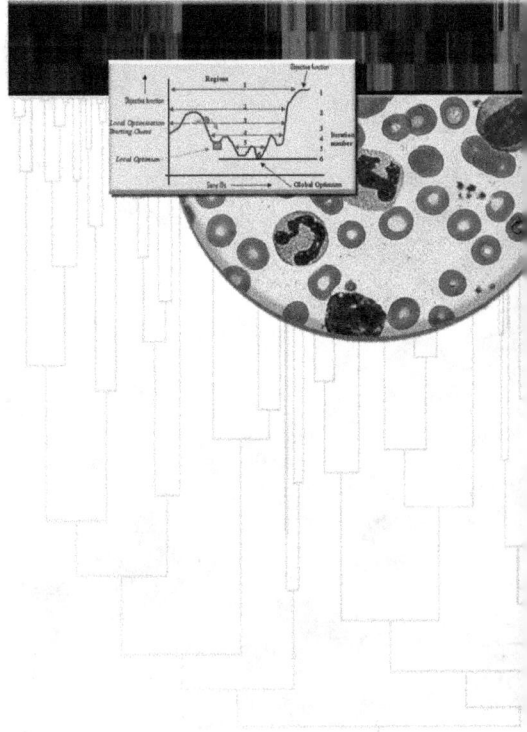

DNA
Microarray Technology and
Data Analysis in Cancer Research

Shaoguang Li

University of Massachusetts Medical School, USA

Dongguang Li

Edith Cowan University, Australia

World Scientific

NEW JERSEY · LONDON · SINGAPORE · BEIJING · SHANGHAI · HONG KONG · TAIPEI · CHENNAI

Published by

World Scientific Publishing Co. Pte. Ltd.

5 Toh Tuck Link, Singapore 596224

USA office: 27 Warren Street, Suite 401-402, Hackensack, NJ 07601

UK office: 57 Shelton Street, Covent Garden, London WC2H 9HE

British Library Cataloguing-in-Publication Data
A catalogue record for this book is available from the British Library.

DNA MICROARRAY TECHNOLOGY AND DATA ANALYSIS IN CANCER RESEARCH

ISBN-13 978-981-279-045-3
ISBN-10 981-279-045-4

Typeset by Stallion Press
Email: enquiries@stallionpress.com

Preface

DNA microarray technology has become a useful technique in gene expression analysis for the development of new diagnostic tools, and for the identification of disease genes and therapeutic targets for human cancers. Appropriate control for DNA microarray experiment and reliable analysis of the array data are the key to performing the assay and utilizing the data correctly. The most difficult challenge has been the lack of a powerful method to analyze the data for all genes (more than 30000) simultaneously and to use the microarray data in a decision-making process.

In this book, we attempt to describe DNA microarray technology and data analysis by pointing out current advantages and disadvantages of the technique and available analytical methods. An important part of the book is that we will include some new ideas and analytical methods based on our own experience in DNA microarray study and analysis. We believe that our new way of interpreting and analyzing the microarray data will bring us closer to success in decision making using the information obtained through DNA microarray technology.

Introduction of Authors

Shaoguang Li, M.D., Ph.D.
Division of Hematology/Oncology
Department of Medicine
University of Massachusetts Medical School
Lazare Research Building #315
364 Plantation Street
Worcester, MA 01605
USA
Tel: +1 508 856 1691
Email: Shaoguang.Li@umassmed.edu

Dr. Shaoguang Li obtained his M.D. degree from China Medical University, Shenyang, China; and his Ph.D. degree from Tulane University, New Orleans, Louisiana, USA. He did his postdoctoral training at Harvard Medical School in Boston, Massachusetts, USA; and began his independent scientific career at The Jackson Laboratory, Bar Harbor, Maine, USA. He is currently an Associate Professor at the University of Massachusetts Medical School, Worcester, Massachusetts, USA. His research interests have been in understanding the molecular basis of leukemogenesis and in developing effective and curative therapeutic strategies. He has made significant contributions to the research field of Philadelphia chromosome-positive (Ph[+]) leukemia using a mouse leukemia model, including the discovery of the roles of Src family kinases and leukemia stem cells in leukemia initiation/progression and resistance to kinase inhibitors. One of his main research focuses is on understanding the biology of leukemia stem cells for developing curative anti-stem cell therapy for Ph[+] leukemia and other types of cancer.

Dongguang Li, Ph.D.
School of Computer and Information Science
Edith Cowan University
2 Bradford Street
Mount Lawley, WA 6050
Australia
Tel: +61 8 93706358
Email: d.li@ecu.edu.au

Dr. Dongguang Li is currently an Associate Professor in the School of Computer and Information Science at Edith Cowan University, Australia, and is a member of the Institute of Electrical and Electronics Engineers (IEEE) and the Optical Society of America (OSA). He obtained an M.Eng. degree in optical engineering from the University of Shanghai for Science and Technology, China, in 1984; and a Ph.D. in physics from the University of New South Wales, Australia, in 1994.

He has had a productive research career spanning more than two decades. During this time, he has made important and significant contributions to a variety of scientific fields, in particular the broad fields of optical engineering, global optimization, and information technology.

His research interests have been in global optimization algorithms and their applications in intelligent ballistics image processing and optical thin film design. He invented a unique and innovative true global optimization algorithm for optical thin film design. He is the author of the optical thin film design software, OpteFilm, and its upgraded version, OnlyFilm. He has also established a database system based on ballistics image processing for firearm identification, for which he was awarded the 2006 Australian B-HERT (Business–Higher Education Round Table) Award for best research and development collaboration. Recently, he has expanded his research interests to the field of microarray data mining.

Contents

CHAPTER 1

DNA Microarray Technology

All living organisms are composed of cells. As a functional unit, each cell can make copies of itself, and this process depends on a proper replication of the genetic material known as deoxyribonucleic acid (DNA). DNA contains genes, and each structural gene functions by transcribing it into the corresponding messenger RNA (mRNA) using DNA as a template and ultimately translating into the corresponding protein using mRNA as a template (Fig. 1.1). The abundance and stability of proteins determine the functions of a cell. Thus, the function or activity of a gene is reflected by synthesis of mRNA (transcription) or protein (translation). DNA microarray technology measures the activity of genes at a transcriptional level.

DNA microarrays (sometimes called DNA chips) are in general characterized by a structured immobilization of DNA targets in the free nucleic acid samples on planar solid supports, on which different types of nucleic acids with known sequences (known as "probes") are fixed. A probe may be derived from complementary DNA (cDNA), polymerase chain reaction (PCR) products, or synthetic oligomers. In general, applications of DNA microarray technology broadly include (1) gene expression analysis (transcription analysis), which analyzes the transcriptional activity of genes through hybridization between DNA targets and probes; (2) genotyping with oligonucleotide arrays, which is based on the notion of combining the complete sequence of a DNA sample by presenting all possible sequences as a complement on the chip (Drmanac *et al.*, 2002); (3) measurement of enzyme activities on immobilized DNA, which is based on the finding that DNA-modifying enzymes are capable of acting on immobilized DNA templates or oligonucleotides (Bier *et al.*, 1996a; Bier *et al.*, 1996b; Buckle *et al.*, 1996); (4) PCR on the chip, which was

1

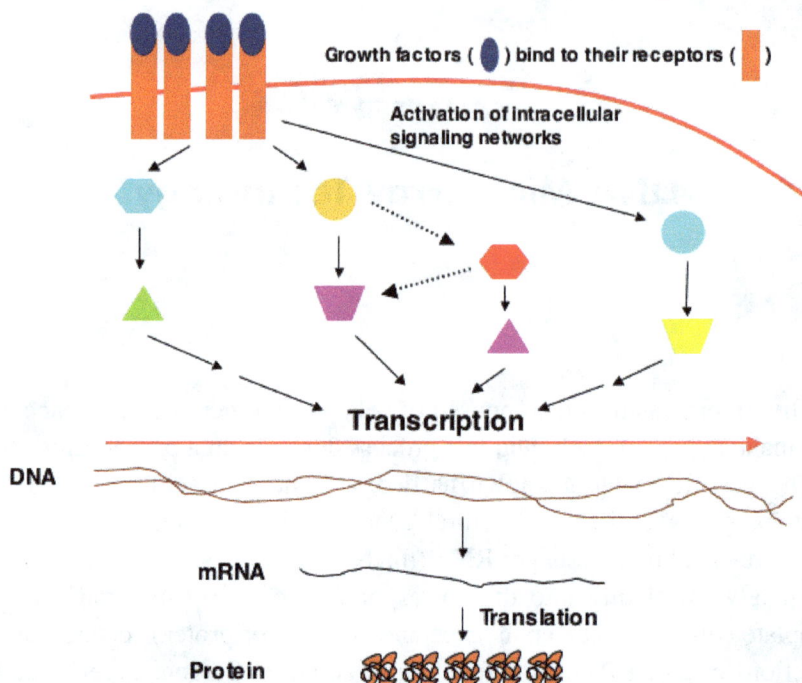

Fig. 1.1.   Activation of signaling pathways at different levels in cells that are stimulated by growth factors. Binding of growth factors to the cell surface receptor leads to activation of intracellular downstream known (solid lines) and unknown (broken lines) signaling molecules (colored blocks). The end result of this activation is to turn on the transcription of genes and ultimately the translation of mRNA into proteins that promote or inhibit cell growth. Measurement of synthesis of either mRNA or protein will reflect the function of a gene.

first described in 2000 (Adessi *et al.*, 2000); and (5) transcription on chip, which shows the transcription of a complete gene into mRNA on the chip (Steffen *et al.*, 2005). In this book, we focus on gene expression analysis using DNA microarray technology.

1.1. Experimental Procedure

The procedure of a DNA microarray experiment includes multiple steps from sample preparation to data analysis (Fig. 1.2), among which

```
┌────────────────────────────────────────────────┐
│ 1.  Sample preparation:                         │
│      - isolation of total RNA                   │
│      - reverse transcription and amplification  │
│      - labeling                                 │
└────────────────────────────────────────────────┘
                        │
                        ▼
┌────────────────────────────────────────────────┐
│ 2.  Hybridization:                              │
│      - binding between the targets and probes   │
│      - washing                                  │
└────────────────────────────────────────────────┘
                        │
                        ▼
┌────────────────────────────────────────────────┐
│ 3.  Detection:                                  │
│      - chip reading                             │
└────────────────────────────────────────────────┘
                        │
                        ▼
┌────────────────────────────────────────────────┐
│ 4.  Data acquisition and analysis:              │
│      - collection and summary of raw data       │
│      - statistical analysis of the data         │
└────────────────────────────────────────────────┘
```

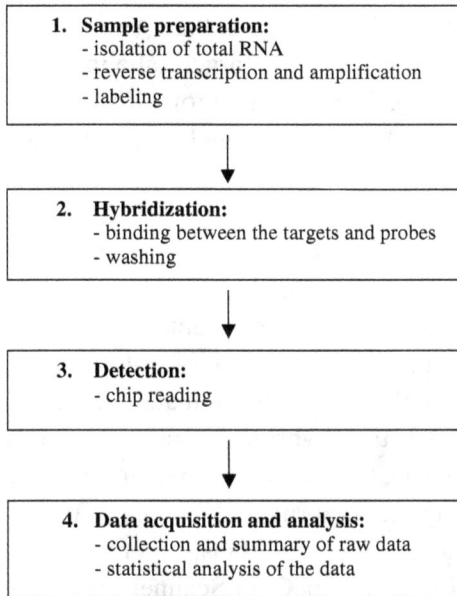

Fig. 1.2.   Workflow of a DNA microarray experiment.

hybridization is a central process. The sample that contains the targets to be investigated is added to the DNA chip to allow their binding to the probes fixed on the chip, resulting in a characteristic-binding pattern representing the levels of gene expression of the sample. The sample itself is labeled prior to the hybridization, and the most-often-used labels are fluorescence labels that allow detection of the binding event. After the hybridization, the chip is washed and then fluorescence intensities on the chip are read and recorded by a scanning or imaging device. Raw data reflecting the fluorescence intensities are statistically analyzed and often shown by fold changes as compared to control.

In our laboratory, we performed DNA microarray experiments to study gene expression profiling in mouse leukemia cells. Here is an example of the procedures for carrying out the microarray experiments. Briefly, cells are dissolved in RNAlater (Ambion, Austin, TX, USA) and homogenized in RLT Buffer (RNeasy Micro Kit; Qiagen, Valencia, CA, USA). Total RNA is isolated by following the protocol for the RNeasy Micro Kit, and quality is assessed using a 2100 Bioanalyzer instrument and RNA 6000 Pico

LabChip assay (Agilent Technologies, Palo Alto, CA, USA). Utilizing the GeneChip Whole Transcript Sense Target Labeling Assay kit (Affymetrix, Santa Clara, CA, USA), 100–300 ng of total RNA undergoes reverse transcription with random hexamers tagged with T7 sequence. The double-stranded cDNA that is generated is then amplified by T7 RNA polymerase to produce cRNA. Second-cycle first-strand cDNA synthesis then takes place, incorporating dUTP, which is later used as sites where fragmentation occurs by utilizing a uracil DNA glycosylase and apurinic/apyrimidinic endonuclease 1 enzyme mix. The fragmented cDNA is then labeled by terminal transferase, attaching a biotin molecule using Affymetrix proprietary DNA Labeling Reagent. Approximately 2.0 µg of fragmented and biotin-labeled cDNA is then hybridized onto a Mouse Gene ST 1.0 Array (Affymetrix, Santa Clara, CA, USA) for 16 hours at 45°C. Posthybridization staining and washing are performed according to the manufacturer's protocols using the Fluidics Station 450 instrument (Affymetrix). Finally, the arrays are scanned with a GeneChip Scanner 3000. Images are acquired and CEL files generated, which are then used for data analysis. Figure 1.3 shows an example of the fluorescence intensities on the chip recorded by an imaging device.

1.2. Experimental Design

Based on our experience, the most critical factor in the experimental design of a DNA microarray experiment is to have a correct control for comparison, which means that we need to appropriately choose a cell or tissue source for isolating control RNA. Choosing a correct control can be extremely challenging, and sometimes it may not be possible to find a "perfect" control. For example, DNA microarray technology has been widely used in comparing gene expression profiling between tumor cells or tissues and corresponding normal cells or tissues in humans. Typically, total RNA is isolated from tumor cells/tissues that are heterogeneous in origin, and control RNA isolated from cells/tissues adjacent to the tumor cells/tissues or corresponding normal cells/tissues is also heterogeneous in most cases; cellular compositions of tumor and normal tissues are not the same or sometimes not even closely similar for appropriate comparisons. Although this cellular difference between tumor and normal tissues is a

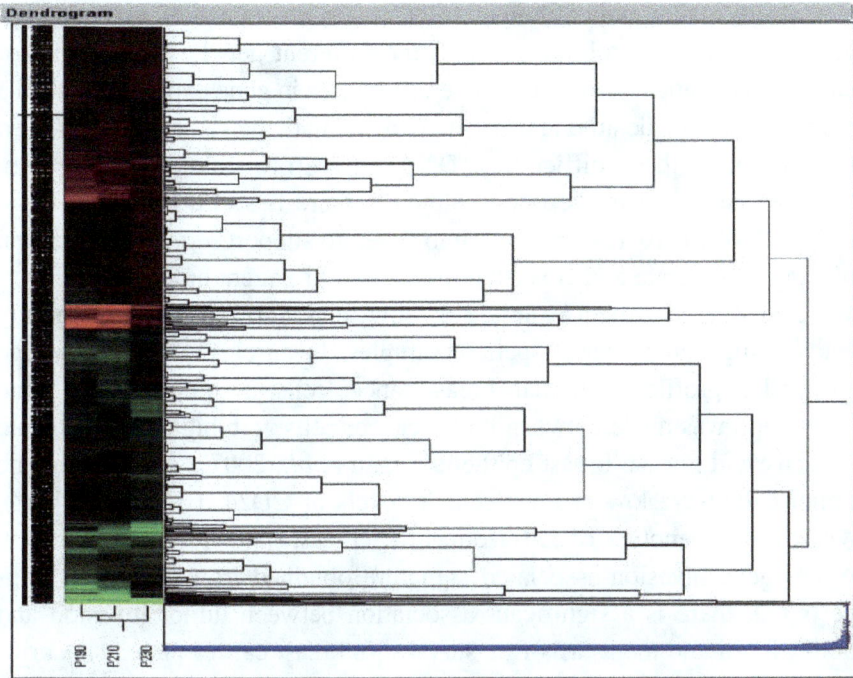

Fig. 1.3. DNA microarray analysis of gene expression profiles in pre-B leukemic cells expressing P190, P210, or P230 BCR-ABL. Total RNA was isolated from the parental ENU pre-B cell line and the same cell line expressing P190, P210, or P230. A total of 24 000 genes were measured using Affymetrix GeneChip, and the data were analyzed by F2 test. Expression of 2086 genes in BCR-ABL-expressing cells was found to be significantly different from that for the reference (parental cell control) ($\alpha < 0.05$). The scientific background of the pre-B leukemic cells expressing P190, P210, or P230 BCR-ABL will be introduced in Chapter 2.

difficult technical problem in microarray studies, the current approach may have done its best to find the most appropriate controls available. Some of the findings obtained from this type of studies are real and useful, although some genes that play critical roles in tumor formation might not be detected due to the inappropriate controls used. A better way to improve this situation is to isolate subpopulations of tumor cells with antibodies recognizing phenotypical cell surface markers. This approach can be taken much more easily when studying human blood cancer cells. However, the availability and amount of human cells for sorting out a particular cell population can be difficult. In parallel, the different genetic background of each human patient will cause some differences in gene expression when a comparison is made among different patients with the same type of cancer. Regardless of these difficulties, DNA microarray technology has been found to be useful in the clinic, although there is still a long way to go before DNA microarray results can be used to support regulatory decision making or accurate and consistent prediction of patient outcomes.

Here is an excellent example of how to perform a DNA microarray study using human cancer patient samples. In an elegant study of gene expression profiles of human breast cancer cells, a comparison of gene expression was made between breast cancer cells with higher tumorigenic capacity and normal breast epithelium (Liu *et al.*, 2007). Specifically, the tumor cells were low or undetectable levels of CD24 (CD44$^+$CD24$^{-/low}$), whereas the phenotype of cells from normal breast epithelium was unknown. A 186-gene invasion-associated signature obtained from this study suggests that there is a significant association between tumor invasion and both overall and metastasis-free survival in breast cancer patients. A critical question to ask is whether the normal breast epithelium is an appropriate control in this DNA microarray study. No matter what the answer is, it is clear that the normal breast epithelium is one of the best controls available.

1.3. Quality Control

A major issue in DNA microarray technology is its repeatability and reproducibility. Repeatability refers to the ability to provide closely similar results from replicate samples processed in parallel at the same test location using the same gene expression assay. Reproducibility refers to the ability to provide closely similar results from replicate samples processed with different microarray platforms or at different test locations using the same gene expression assay. To achieve high repeatability and reproducibility, quality control has become a key issue in DNA microarray studies.

A major criticism voiced about DNA microarray studies has been the lack of accuracy and reproducibility of the microarray data. The quality of DNA microarray results is associated with technical, instrumental, computational, and interpretative factors (Casciano and Woodcock, 2006). For the same samples tested, results obtained at different locations and between different microarray platforms could be different, making it difficult to use and interpret microarray data. Optimization and standardization of microarray procedures are critical steps. In this regard, the US Food and Drug Administration (FDA) has initiated a MicroArray Quality Control (MAQC) project among researchers in academic, government, and industrial institutions to seek to experimentally address the key issues surrounding the reliability of DNA microarray data. As part of this effort, since 2004 the FDA has started accepting voluntary genomic data submission with accompanying information related to the number and scope of DNA microarray-based expression data (Anonymous, 2006). The MAQC project aims to establish quality control metrics and thresholds for an objective assessment of the performance achievable by different microarray platforms, and for evaluation of the merits and limitations of various data analysis methods (Casciano and Woodcock, 2006). The specific concerns of the MAQC project relate to the impact of microarray data quality on genomic data submission (Frueh, 2006; Ji and Davis, 2006), the framework for the use of genomics data (Dix *et al.*, 2006), comparisons of different commercial microarray platforms (Canales *et al.*, 2006; Patterson *et al.*, 2006; Shippy *et al.*, 2006), the use of external RNA controls (Tong *et al.*, 2006), interplatform and intraplatform reproducibility of gene expression measurements (Shi *et al.*, 2006), and analytical consistency across microarray platforms (Guo *et al.*, 2006).

The main conclusion of the MAQC project is that, with careful experimental design and appropriate data transformation and analysis, microarray data can indeed be reproducible and comparable between different formats and laboratories, and that fold change results from microarray experiments correlate closely with results from well-accepted assays such as quantitative reverse transcription–polymerase chain reaction (qRT-PCR) (Anonymous, 2006). However, there is still a long way to go before we can answer the key remaining question: when can microarray data be used in a regulatory decision-making process?

1.4. Interpretation of DNA Microarray Data

While expensive, DNA microarray experiments have not yielded sufficient information that allows us to draw decisive conclusions. Often, what we have in the end is a long list of more than 30 000 genes with fold changes of their mRNA expression comparing control and experimental groups. An easy explanation for the fold change of mRNA expression of a particular gene, which is actually a way of obtaining information from DNA microarray data by most people, is that mostly changed genes in expression are likely involved in the biological process being studied. A "hot" list is often provided by statisticians to researchers to describe which genes are mostly upregulated or downregulated, and those genes with minor or no changes in mRNA expression are often considered as not involved or not important by the researchers. However, there are numerous examples showing a disassociation between the abundance of mRNA and the level of translated protein for a gene of interest, and between the abundance of mRNA and the effect of the gene of interest on a particular biological process.

From our own experience, a general guess on the biological role of a particular gene based on the fold change is sometimes agreeable to the results of experiments testing the function of this gene, but it is not rare to see misleading or wrong conclusions drawn based solely on the magnitude of the fold changes. Importantly, we learned that upregulation or downregulation of a gene does not necessarily mean a positive or negative role of this gene in the biological process being studied. A novel idea we intend to propose here is that a gene with no change in its expression, as determined by DNA microarray analysis, may still play a significant role in the biological process. In Chapter 2, we describe in much more detail and emphasize these ideas about the interpretation of DNA microarray data, with specific examples from our own studies.

1.5. Advantages and Disadvantages

One of the biggest advantages of DNA microarray technology is that it can evaluate simultaneously the relative expression of thousands of genes by using small amounts of materials, providing gene signatures for particular disease situations. In addition, the procedures can easily be automated. Furthermore, the capacity of measurement of gene expression by DNA microarray is huge, allowing researchers to take the expression of all genes from an individual into consideration for disease analysis in so-called "personalized medicine".

One of the major disadvantages of DNA microarray technology is that it only evaluates gene expression at a transcriptional, but not translational, level, as posttranscriptional modifications (such as phosphorylation) often play significant roles in the regulation of protein functions. In addition, DNA microarray technology is still not mature enough for decision making based on the microarray data. In this book, we introduce our new way of interpreting and analyzing microarray data, which will hopefully bring us closer to success in decision making using the information obtained through DNA microarray technology (see Chapter 4).

CHAPTER 2

Applications of DNA Microarray Technology in Cancer Research

Seven years ago, the sequencing of the human genome was completed (Lander *et al.*, 2001; Venter *et al.*, 2001), and many discussions and comments about the value of this information have been initiated since then. However, it is obvious that DNA microarray technology would not be possible without knowing the genome sequences of humans, mice, and other species as well. This technology has been widely used in basic and clinical studies of virtually all major human diseases, especially in cancer research. There are many good examples for studying gene expression in cancers, and here we only discuss some representative studies to show the applications of DNA microarray technology in cancer research.

2.1. Solid Tumors

A DNA microarray study from Pantel's group showed molecular signature associated with bone marrow micrometastasis in human breast cancer (Woelfle *et al.*, 2003). In this study, gene expression profiles in metastasized breast tumor cells in bone marrow were compared with those in primary tumor cells, and expression analysis showed distinct profiles between these two groups of cells. The differentially expressed genes are related to extracellular matrix remodeling, adhesion, cytoskeleton plasticity, and signal transduction (in particular, the Ras and hypoxia-inducible factor 1α pathways). The array data were confirmed by RT-PCR, which is consistent with immunohistochemical analysis of breast tumor tissues. The findings from this study indicate that metastasized breast tumor cells exist as a selective process associated with a specific molecular signature. Study of the functional relevance of this molecular signature will shed light on the molecular diagnosis and therapy of human breast cancer.

Another study published by Di Fiore's group illustrated a beautiful experiment to show that survival of stage 1 lung adenocarcinomas could be predicted based on the expression of 10 genes discovered through analyzing the clinical DNA microarray data (Bianchi *et al.*, 2007). The beauty of this study is that the original clues for identifying candidate genes came from their analyses of two pre-existing sets of clinical microarray data from two independent places (Michigan and Harvard, USA). The original datasets from Michigan included 86 adenocarcinomas of human lung cancer; those from Harvard included 84 cases. By analyzing these microarray data, a 80-gene model was created and tested on an independent cohort of lung cancer patients using RT-PCR. As a result, a 10-gene predictive model exhibited a prognostic accuracy of approximately 75% in stage 1 lung adenocarcinoma when tested on two additional independent cohorts. Potentially, this type of approach enables the discovery of similar predictive molecular signatures for other types of cancer.

There are many more examples showing the application of DNA microarray technology in identifying molecular targets and establishing diagnostic molecular signatures for human cancers. The overall weakness of these microarray studies is the lack of ability to make decisions based solely on the DNA microarray data.

2.2. Blood Cancers

An excellent example for using DNA microarray technology to study human blood cancers is the molecular classification of human acute myeloid leukemia (AML) and acute lymphoblastic leukemia (ALL) (Golub *et al.*, 1999). In this study, bone marrow mononuclear cells from 11 AML and 27 ALL patients diagnosed pathologically were used as an RNA source for DNA microarray analysis, from which 50-gene predictors that distinguish AML from ALL were derived. These 50-gene predictors were tested and validated on 38 new samples from AML or ALL patients. As a result, 36 of the 38 predictions agreed with the patients' clinical diagnosis (the remaining two were uncertain). This high prediction rate (95%)

strongly suggests that DNA microarray technology may be used in the diagnosis of human blood cancers, although the improvement to a 100% prediction rate is the ultimate goal.

2.3. Our DNA Microarray Study Using Mouse Model of BCR-ABL-Induced Leukemia

Here, we use our DNA microarray studies as examples to show our views on how a microarray experiment should be designed and how the data should be interpreted. To help explain the rationale of our study on leukemia cells and the mouse leukemia model we used, we first introduce some disease-related background information.

Human Philadelphia chromosome-positive (Ph⁺) leukemias induced by the BCR-ABL oncogene are among the most common hematologic malignancies, and include chronic myeloid leukemia (CML) and B-cell ALL (B-ALL). The BCR gene, on chromosome 22, breaks at either exon 1, exon 12/13, or exon 19; and fuses to the c-ABL gene on chromosome 9 to form, respectively, three types of BCR-ABL: P190, P210, or P230. In humans, each of the three forms of the BCR-ABL oncogene is associated with a distinct type of leukemia [see review by Advani and Pendergast (2002)]. The P190 form is most often present in B-ALL, but only rarely in CML. P210 is the predominate form in CML, and in some acute lymphoid and myeloid leukemias in CML blast crisis. P230 was recently found in a very mild form of CML (Pane *et al.*, 1996). CML has a triphasic clinical course: a chronic phase, in which BCR-ABL-expressing pluripotent stem cells massively expand but undergo normal differentiation to form mature neutrophils; an accelerated phase, in which neutrophil differentiation becomes progressively impaired and the cells become less sensitive to myelosuppressive medications; and blast crisis, a condition resembling acute leukemia in which myeloid or lymphoid blasts fail to differentiate. The transition from chronic phase to blast crisis results from additional genetic alterations, and this process is not well understood. Lymphoid blast crisis of CML and Ph⁺ B-ALL account for 20% of adults and 5% of children afflicted with ALL. Among those patients, 50% of adults and 20% of children carry P210BCR-ABL while the rest carry P190BCR-ABL (Druker *et al.*, 2001a; Sawyers, 1999).

The Abl tyrosine kinase inhibitor STI571 (imatinib mesylate; Gleevec) has become the most effective drug for leukemia therapy, and has been shown to induce a complete hematologic response in all interferon-resistant chronic-phase CML patients (Druker *et al.*, 2001b). However, STI571 was unable to abrogate BCR-ABL-expressing leukemic cells (Marley *et al.*, 2000) and induced cellular and clinical drug resistance (Branford *et al.*, 2002; Gorre *et al.*, 2001; le Coutre *et al.*, 2000; Mahon *et al.*, 2000; Shah *et al.*, 2002; von Bubnoff *et al.*, 2002; Weisberg and Griffin, 2000), suggesting that the use of STI571 as a single agent may not prevent eventual disease progression to terminal blast crisis or cure CML. Moreover, STI571 is much less effective in treating CML blast crisis patients (Druker *et al.*, 2001a; Talpaz *et al.*, 2000). Therefore, the development of new therapeutic drugs that are synergistic with available treatment strategies is critical.

The success of this approach requires an understanding of the signaling pathways utilized by BCR-ABL to induce CML and B-ALL. Expression of BCR-ABL has been shown to activate multiple signaling molecules/ pathways, including Ras, MAPK, STAT, JNK/SAPK, PI-3 kinase, NF-κB, and c-MYC (Sawyers, 1997), as well as cytokine production (Anderson and Mladenovic, 1996; Hariharan *et al.*, 1988). Recent studies also link BCR-ABL to apoptotic pathways (Amarante-Mendes *et al.*, 1998; Dubrez *et al.*, 1998; Goetz *et al.*, 2001; Honda and Hirai, 2001; Jonuleit *et al.*, 2000; Majewski *et al.*, 1999; McGahon *et al.*, 1995; Neshat *et al.*, 2000; Parada *et al.*, 2001; Sanchez-Garcia and Martin-Zanca, 1997; Skorski *et al.*, 1997; Skorski *et al.*, 1996) and to the activation of Src kinases in cultured cells (Danhauser-Riedl *et al.*, 1996; Lionberger *et al.*, 2000; Warmuth *et al.*, 1997). Our studies in mice have shown the involvement of three Src family kinases — Lyn, Hck, and Fgr — in BCR-ABL-induced B-lymphoid leukemia (Hu *et al.*, 2004). It is important to point out that, although these signaling molecules are present and activated in BCR-ABL-expressing cells, it remains unclear what role, if any, they play in the development of BCR-ABL-induced leukemias. The physiological relevance of the *in vitro* findings of the roles of signaling molecules in BCR-ABL signaling is better addressed *in vivo* using the mouse leukemia model, although *in vitro* studies (such as the use of cultured cells or cell lines) are often useful in finding preliminary clues.

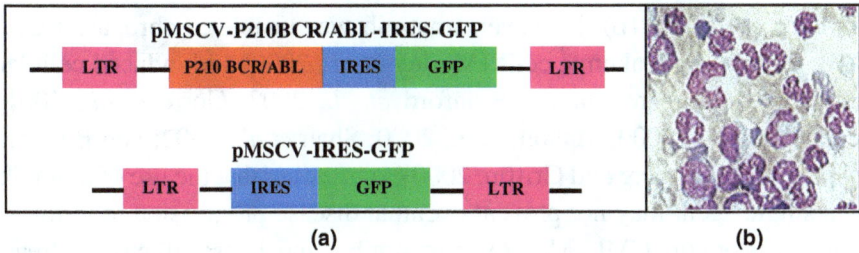

Fig. 2.1.   BCR-ABL induces both myeloid (CML) and lymphoid (B-ALL) diseases in mice. (a) The retroviral constructs used to transduce the P210BCR-ABL and GFP genes. (b) Wright/Giemsa staining of peripheral blood smears from mice with CML.

Although BCR-ABL induces both CML and B-ALL, here we only focus on the induction of CML by BCR-ABL. By retrovirally expressing the BCR-ABL oncogene in bone marrow cells, which are derived from donor BALB/c mice pretreated with 5-fluorouracil (5-FU), we are now able to produce CML in 100% of syngeneic recipients within 4 weeks (Li *et al.*, 1999). The same CML can also be induced in C57BL/6 (B6) and other inbred strains. Mouse CML is characterized by leukocytosis with greatly elevated numbers of maturing neutrophils [Fig. 2.1(a)], organ infiltration, and splenomegaly. The target cells for BCR-ABL are primitive multipotential hematopoietic stem cells (Li *et al.*, 1999), precisely mimicking most of the pathological characteristics of human CML. Myeloid cells transduced by the BCR-ABL oncogene express BCR-ABL protein (Li *et al.*, 1999). The control retroviral vector containing only the GFP gene [Fig. 2.1(b)] has never induced CML in recipient mice within 12 months (data not shown). We carried out a microarray study to identify genes that are regulated by BCR-ABL in cell lines and in our mouse CML model.

2.3.1. Leukemia mouse model study

2.3.1.1. Rationale

The Abl tyrosine kinase inhibitor imatinib has become the most effective drug for leukemia therapy, and has been shown to induce a complete hematologic response in all interferon-resistant chronic-phase CML patients

(Druker *et al.*, 2001b). However, imatinib was unable to abrogate BCR-ABL-expressing leukemic cells (Marley *et al.*, 2000) and induced cellular and clinical drug resistance (Branford *et al.*, 2002; Gorre *et al.*, 2001; le Coutre *et al.*, 2000; Mahon *et al.*, 2000; Shah *et al.*, 2002; von Bubnoff *et al.*, 2002; Weisberg and Griffin, 2000), suggesting that the use of imatinib as a single agent may not prevent eventual disease progression to terminal blast crisis or cure CML. Moreover, imatinib is much less effective in treating CML blast crisis patients (Druker *et al.*, 2001a; Talpaz *et al.*, 2000).

Recently, a newly developed Abl kinase inhibitor (termed BMS-354825 and produced by Bristol-Myers Squibb) has been shown to have an inhibitory effect on almost all imatinib-resistant BCR-ABL mutants (Shah *et al.*, 2004), offering some hope for overcoming imatinib resistance. However, the BCR-ABL-T315I mutant that frequently appears in patients resistant to imatinib therapy is still resistant to BMS-354825 (Shah *et al.*, 2004). Another novel Abl kinase inhibitor, named AP23464, has also been effective against several frequently observed imatinib-resistant BCR-ABL mutants, but ineffective against the BCR-ABL-T315I mutant (O'Hare *et al.*, 2004). While clinical trials are ongoing to determine the effectiveness of these drugs in treating human Ph[+] leukemia patients who are resistant to imatinib, our preliminary results have shown that imatinib does not eradicate leukemic cells in mice. Therefore, it is critical to develop new therapies that are synergistic with available treatment strategies and capable of killing BCR-ABL-expressing leukemic stem cells (LSCs). The success of this approach requires an understanding of the signaling pathways utilized by BCR-ABL in LSCs to induce Ph[+] leukemias. Molecular pathways involved in regulation of the self-renewal and survival of Ph[+] LSCs are largely unknown. We have identified BCR-ABL-expressing hematopoietic stem cells (HSCs) (GFP[+]Lin[-]c-Kit[+]Sca-1[+]) as CML stem cells (Hu *et al.*, 2006), providing a valuable assay system for studying the biology of LSCs. Using this LSC assay system, we will be able to identify and test molecular targets in LSCs.

2.3.1.2. *Experimental design*

Bone marrow cells are isolated from the long bones of CML mice that are untreated or treated with imatinib. BCR-ABL-expressing or non-BCR-ABL-expressing (transduced with the empty GFP vector) HSCs (GFP[+]Lin[-]c-Kit[+]Sca-1[+]) are stored by fluorescence-activated cell sorting

(FACS) directly into RNAlater (Ambion, Austin, TX, USA) and are homogenized in RLT Buffer (RNeasy Micro Kit; Qiagen, Valencia, CA, USA). Total RNA is isolated by following the protocol for the RNeasy Micro Kit, and quality is assessed using a 2100 Bioanalyzer instrument and RNA 6000 Pico LabChip assay (Agilent Technologies, Palo Alto, CA, USA). Utilizing the GeneChip Whole Transcript Sense Target Labeling Assay kit (Affymetrix, Santa Clara, CA, USA), 100–300 ng of total RNA undergoes reverse transcription with random hexamers tagged with T7 sequence. The double-stranded cDNA that is generated is then amplified by T7 RNA polymerase to produce cRNA. Second-cycle first-strand cDNA synthesis then takes place, incorporating dUTP, which is later used as sites where fragmentation occurs by utilizing a uracil DNA glycosylase and apurinic/apyrimidinic endonuclease 1 enzyme mix. The fragmented cDNA is then labeled by terminal transferase, attaching a biotin molecule using Affymetrix proprietary DNA Labeling Reagent. Approximately 2.0 µg of fragmented and biotin-labeled cDNA is then hybridized onto a Mouse Gene ST 1.0 Array (Affymetrix, Santa Clara, CA, USA) for 16 hours at 45°C. Posthybridization staining and washing are performed according to the manufacturer's protocols using the Fluidics Station 450 instrument (Affymetrix). Finally, the arrays are scanned with a GeneChip Scanner 3000. Images are acquired and CEL files generated, which are then used for analysis.

2.3.1.3. *Results*

The microarray data — shown as upregulation, downregulation, or no change — of more than 30 000 mouse genes in the presence and absence of BCR-ABL in LSCs have been sent to the National Center for Biotechnology Information (NCBI) through Gene Expression Omnibus (GEO) submission. In sum, several gene expression patterns are observed when considering how the expression of a particular gene is changed before and after the treatment with the BCR-ABL kinase inhibitor imatinib (Table 2.1):

- Expression of a gene is upregulated by BCR-ABL as compared to the control, and imatinib treatment reduces the expression level of the gene to the control level (Pattern A).

Table 2.1.   Gene expression patterns of BCR-ABL-expressing cells treated with or without imatinib (IM).

Expression pattern	Gene expression levels in experimental groups			Explanation
	Control	BCR-ABL	BCR-ABL + IM	
A	+++	+++++	+++	positively involved
B	+++	+	+++	negatively involved
C	+++	+++++	+++++ or ++++	positively involved
D	+++	+	+ or ++	negatively involved
E	+++	+++++	+++++++	negatively involved
F	+++	++	+	positively involved
G	+++	+++	+	positively involved
H	+++	+++	+++++	negatively involved
I	+++	+++	+++	not involved

- Expression of a gene is downregulated by BCR-ABL as compared to the control, and imatinib treatment increases the expression level of the gene toward or to the control level (Pattern B).
- Expression of a gene is upregulated by BCR-ABL as compared to the control, and imatinib treatment does not reduce the expression level of the gene toward or to the control level (Pattern C).
- Expression of a gene is downregulated by BCR-ABL as compared to the control, and imatinib treatment does not increase the expression level of the gene toward or to the control level (Pattern D).
- Expression of a gene is upregulated by BCR-ABL as compared to the control, and imatinib treatment further increases the expression level of the gene (Pattern E).
- Expression of a gene is downregulated by BCR-ABL as compared to the control, and imatinib treatment further reduces the expression level of the gene (Pattern F).
- Expression of a gene is not changed by BCR-ABL as compared to the control, and imatinib treatment reduces the expression level of the gene (Pattern G).

- Expression of a gene is not changed by BCR-ABL as compared to the control, and imatinib treatment increases the expression level of the gene (Pattern H).
- Expression of a gene is not changed by BCR-ABL as compared to the control, and the expression level of the gene remains the same upon imatinib treatment (Pattern I).

Detailed explanations for these patterns are described in Table 2.1. In summary, these patterns have changed our way of interpreting DNA microarray data. What we learned from this study is that, in oncogene-stimulated cell proliferation, an upregulated gene is not always positively involved in cancer development, a downregulated gene is not always negatively involved in cancer development, and, importantly, a gene with no change in expression may also be critically involved in cancer development. Thus, the expression level of a gene is relative and does not always reflect its function. It could be misleading to determine the function of a gene based on the expression level of the gene.

2.3.2. Cell line study

The abovementioned explanations of our microarray data are largely different from those most people would have. We tested these ideas by performing an independent microarray experiment in a totally different assay system. We studied gene expression profiling in BCR-ABL-expressing lymphoid cells before and after treatment with imatinib. We first made a B-lymphoid cell line whose growth does not depend on BCR-ABL, and the control cell line is the parental line that does not express BCR-ABL (Hu *et al.*, 2004). Our results also show the gene expression patterns A–J described above, allowing us to reach the same conclusions that an upregulation, downregulation, or no change of expression of a gene does not necessarily reflect the role of this gene in tumor cell growth. A clear conclusion can be drawn by having a critical control that represents a situation when the function of the oncogene of interest is inhibited genetically or by an inhibitory agent.

CHAPTER 3

Current Analytical Methods of DNA Microarray Data

Analysis of DNA microarray data includes multiple steps. Here, we describe some general procedures for the analysis, using our microarray data discussed in Table 3.1 as an example.

3.1. Experimental Design

This experiment had one factor, *Strain*, a fixed factor with three levels (*B6*, *Mut A*, and *Vector* with 3, 2, and 2 biological replicates, respectively). A total of eight Affymetrix Mouse Gene 1.0 ST Arrays were used for gene expression analysis (Table 3.1).

3.2. Method

3.2.1. *Robust multi-chip averaging (RMA)*

Average signal intensities for each probe set within arrays were calculated by and exported from Affymetrix's Expression Console (Version 1.1) software using the RMA method, which incorporates convolution background correction and a summarization based on a multi-array model fit robustly using the median polish algorithm. These values were read into R/maanova and quantile-normalized. For this experiment, three pairwise comparisons will be used to statistically resolve gene expression differences between strain groups using the R/maanova analysis package (Wu *et al.*, 2003). Specifically, differentially expressed genes will be detected by using F_s, a modified F-statistic incorporating shrinkage estimates of

Table 3.1.   Experimental design. Note that Mut_A_1 or Mut_A_2, B6_1 or B6_2, and Vector_1 or Vector_2 are names of the RNA samples for DNA microarray experiment.

Array	Strain	Strain_Rep	Sample	Dye
GC_Gene1.0ST_GES08_0064_040208_1	Mut_A	Mut_A_1	1	1
GC_Gene1.0ST_GES08_0065_040208_1	B6	B6_1	2	1
GC_Gene1.0ST_GES08_0066_040208_1	Vector	Vector_1	3	1
GC_Gene1.0ST_GES08_0067_040208_1	B6	B6_2	4	1
GC_Gene1.0ST_GES08_0068_040208_1	Mut_A	Mut_A_2	5	1
GC_Gene1.0ST_GES08_0069_040208_1	B6	B6_3	6	1
GC_Gene1.0ST_GES08_0070_040208_1	Vector	Vector_2	7	1
GC_Gene1.0ST_GES08_0071_040208_1	B6_P210	B6_P210_1	8	1

variance components from within the R/maanova package (Cui *et al.*, 2005; Wu *et al.*, 2003). Statistical significance levels of the pairwise comparison will be calculated by permutation analysis (1000 permutations) and adjusted for multiple testing using the false discovery rate (FDR), *q*-value, method (Storey, 2002). Differentially expressed genes are declared at an FDR *q*-value threshold of 0.05.

3.2.2.   *iterPLIER*

Average signal intensities for each probe set within arrays were calculated by and exported from Affymetrix's Expression Console (Version 1.1) software using the iterPLIER method with background adjustment (PM-GCBG). These values were read into R/maanova; a value of 16 was added to the processed probeset intensities to account for iterPLIER's zero-based method, log-base-2-transformed, and quantile-normalized to equalize the distribution of intensities across all arrays. For this experiment, three pairwise comparisons will be used to statistically resolve gene expression differences between strain groups using the R/maanova analysis package (Wu *et al.*, 2003). Specifically, differentially expressed genes will be detected by using F_s, a modified *F*-statistic incorporating shrinkage estimates of variance components from within the R/maanova package (Cui *et al.*, 2005; Wu *et al.*, 2003). Statistical significance levels

Table 3.2. **Individual probes with an intensity reading exceeding that of the saturation of the scanner (> 60 000).**

ID	Probe	Row	Intensity
GC_Gene1.0ST_GES08_0065_040208_1			
10387588	12	165140	65533
10599607	32	785261	65534

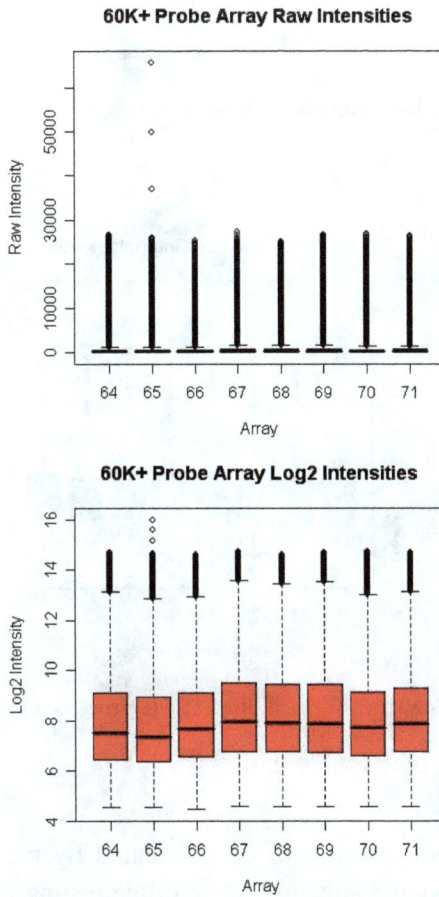

60K+ Probe Array Raw Intensities

60K+ Probe Array Log2 Intensities

Fig. 3.1. **Boxplots of raw (top) and log$_2$-transformed (bottom) intensities for all probes on all arrays.**

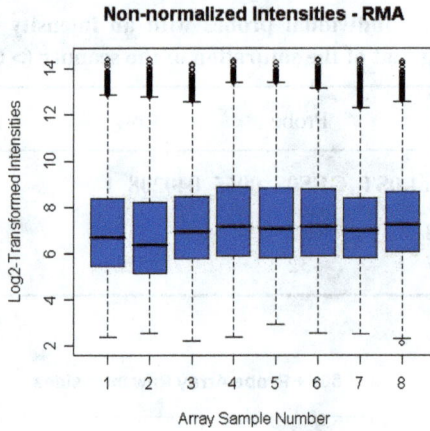

Fig. 3.2. **Boxplot of log$_2$-transformed, RMA-processed probe intensities for all samples.**

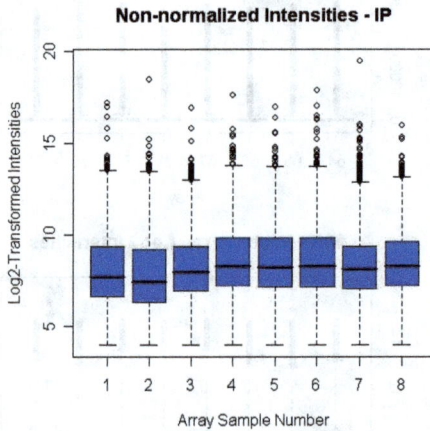

Fig. 3.3.   **Boxplot of log$_2$-transformed, iterPLIER-processed probe intensities for all samples.**

of the pairwise comparison will be calculated by permutation analysis (1000 permutations) and adjusted for multiple testing using the false discovery rate (FDR), q-value, method (Storey, 2002). Differentially expressed genes are declared at an FDR q-value threshold of 0.05.

Normalized Intensities - RMA

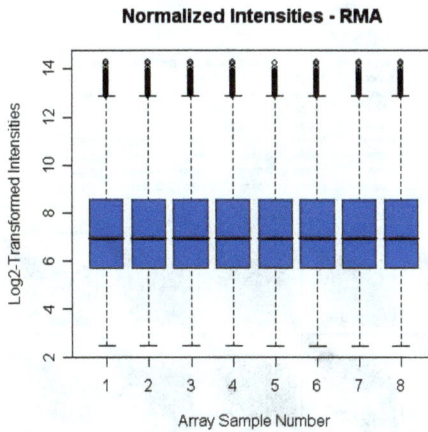

Fig. 3.4.   Boxplot of log$_2$-transformed, quantile-normalized, RMA-processed signal intensities for all samples.

Normalized Intensities - IP

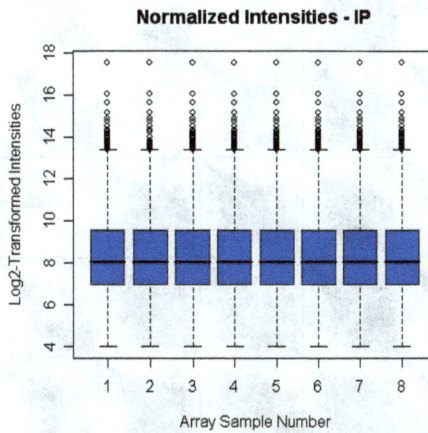

Fig. 3.5.   Boxplot of log$_2$-transformed, quantile-normalized, iterPLIER-processed signal intensities for all samples.

3.3. Quality Control Diagnostics

3.3.1. *Saturation*

Raw intensities for each sample were examined to determine whether any probes had a signal close to the saturation of the scanner

Multivariate

Correlations

	Mut_A_1 (64)	Mut_A_2 (68)	Vector_1 (66)	Vector_2 (70)	B6_1 (65)	B6_2 (67)	B6_3 (69)	B6_P210 (71)
Mut_A_1 (64)	1.0000	0.9431	0.9300	0.9109	0.9009	0.9503	0.9411	0.9400
Mut_A_2 (68)	0.9431	1.0000	0.9463	0.9222	0.9386	0.9662	0.9506	0.9456
Vector_1 (66)	0.9300	0.9463	1.0000	0.9274	0.9024	0.9530	0.9390	0.9413
Vector_2 (70)	0.9169	0.9222	0.9274	1.0000	0.9116	0.9270	0.9199	0.9233
B6_1 (65)	0.9009	0.9300	0.9024	0.9110	1.0000	0.9449	0.9309	0.9372
B6_2 (67)	0.9503	0.9662	0.9536	0.9270	0.9449	1.0000	0.9549	0.9519
B6_3 (69)	0.9411	0.9506	0.9390	0.9199	0.9380	0.9540	1.0000	0.9440
B6_P210 (71)	0.9400	0.9458	0.9413	0.9233	0.9372	0.9519	0.9449	1.0000

Scatterplot Matrix

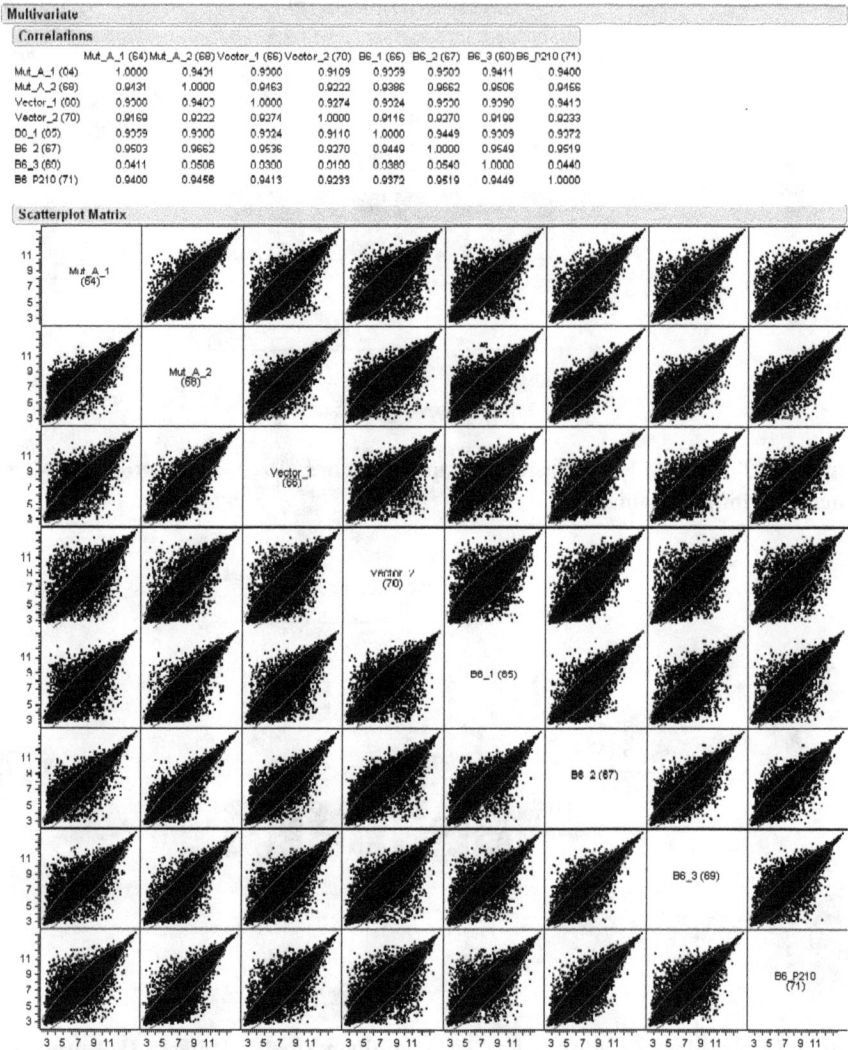

Fig. 3.6. Scatterplots of RMA-processed, normalized intensities for each sample against all other samples.

(>60000). The listed probes had measurements with an intensity reading greater than 60000. The number of probes for all arrays was 6552328. The range of intensities was from 22 to 65534 (Table 3.2).

Multivariate

Correlations

	Mut_A_1 (64)	Mut_A_2 (68)	Vector_1 (66)	Vector_2 (70)	B6_1 (65)	B6_2 (67)	B6_3 (69)	B6_P210_1 (71)
Mut_A_1 (64)	1.0000	0.8814	0.8612	0.8063	0.8668	0.8931	0.8744	0.8664
Mut_A_2 (68)	0.8814	1.0000	0.8824	0.8090	0.8752	0.9309	0.8926	0.8755
Vector_1 (66)	0.8612	0.8824	1.0000	0.8218	0.8522	0.8940	0.8635	0.8622
Vector_2 (70)	0.8063	0.8090	0.8218	1.0000	0.7968	0.8153	0.8091	0.8167
B6_1 (65)	0.8668	0.8752	0.8522	0.7968	1.0000	0.8815	0.8701	0.8576
B6_2 (67)	0.8931	0.9309	0.8940	0.8153	0.8815	1.0000	0.8991	0.8855
B6_3 (69)	0.8744	0.8926	0.8635	0.8091	0.8701	0.8991	1.0000	0.8717
B6_P210_1 (71)	0.8664	0.8755	0.8622	0.8167	0.8576	0.8855	0.8717	1.0000

Scatterplot Matrix

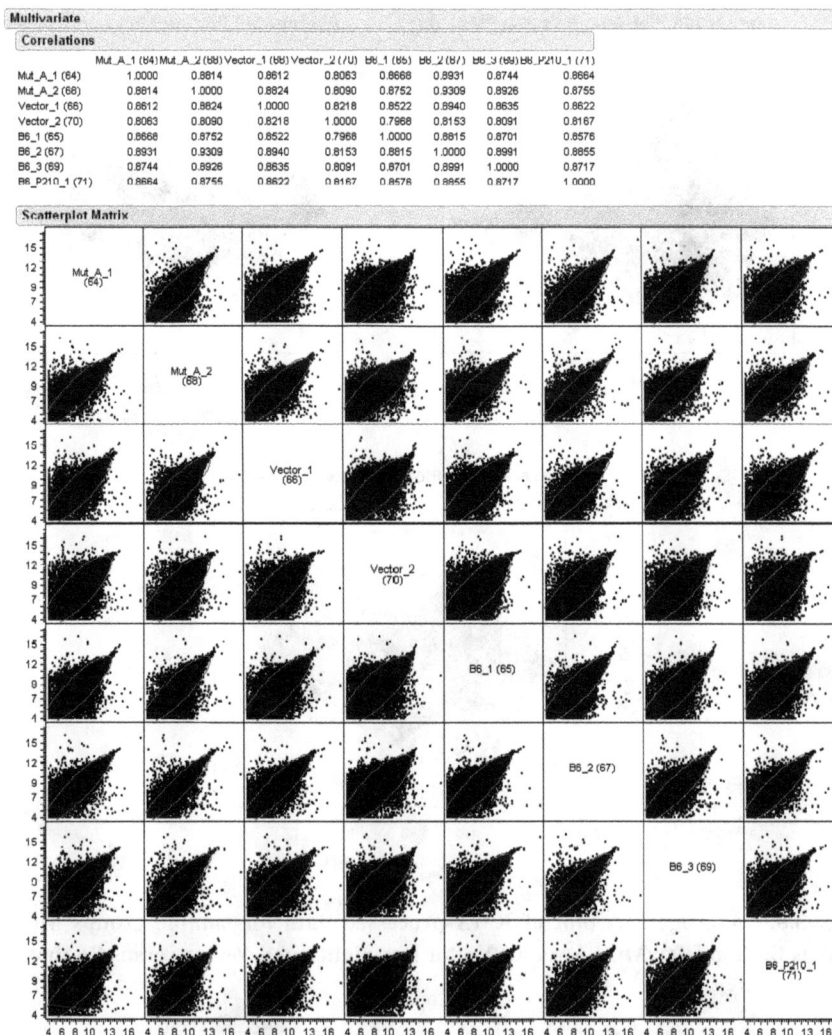

Fig. 3.7. Scatterplots of iterPLIER-processed, normalized intensities for each sample against all other samples.

3.3.1.1. *Raw intensities for arrays containing a saturated probe*

A boxplot of raw and $\log_2$-transformed intensities for all probes for arrays containing a probe with an intensity reading >60000 reveals whether a probe is an outlier or if an array has consistently high readings (Fig. 3.1).

Average MA Plot -Mut_A vs. B6- RMA

Average MA Plot -Mut_A vs. Vector- RMA

Average MA Plot -Vector vs. B6- RMA

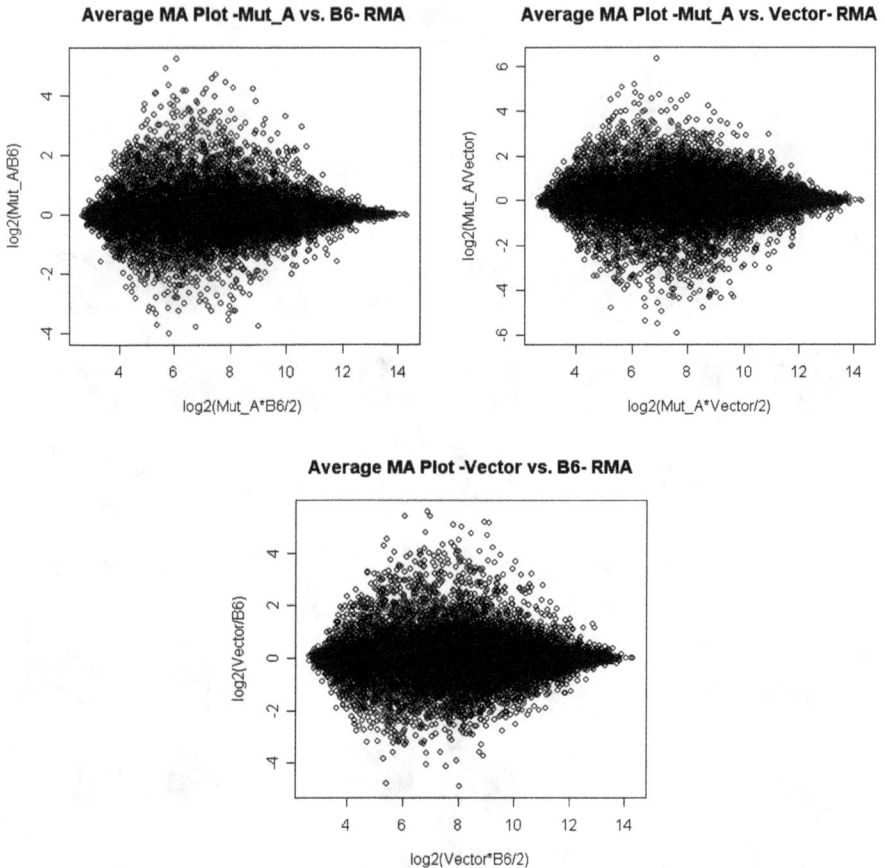

Fig. 3.8. **Average MA plot of RMA-processed data for sample groups Mut_A versus B6 (top left), Mut_A versus Vector (top right), and Vector versus B6 (bottom center).**

3.3.2. *Transformed intensities across arrays*

For each probe set, the processed raw intensities for all probes were averaged and $\log_2$-transformed without normalization from Affymetrix's Expression Console software. The $\log_2$-transformation of raw data was used to reduce the impact of outliers and ensure normality (Figs. 3.2 and 3.3).

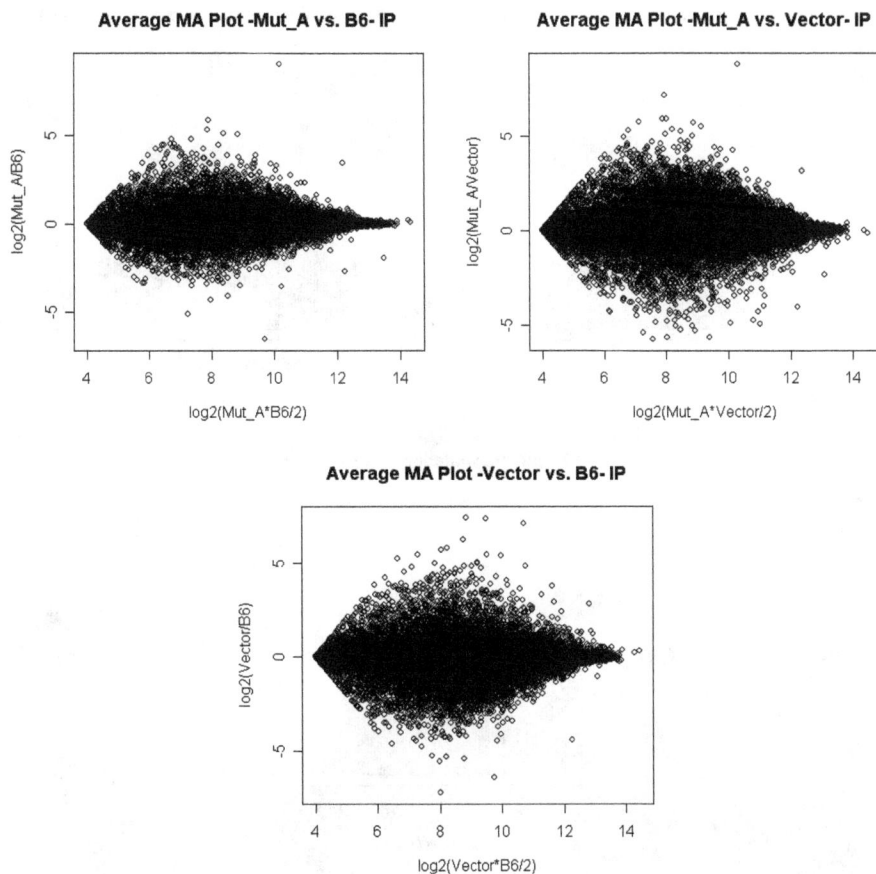

Fig. 3.9. Average MA plot of iterPLIER-processed data for sample groups Mut_A versus B6 (top left), Mut_A versus Vector (top right), and Vector versus B6 (bottom center).

3.3.3. *Normalized intensities across arrays*

The $\log_2$-transformed intensities were quantile-normalized. Quantile-normalization normalizes data based on the magnitude of the measures of probe intensities. The method is employed to remove systematic effects, and to bring the data from different microarrays onto a common scale (Figs. 3.4 and 3.5).

Fs Pvalue Permutation -Mut_A_B6- RMA

Fs Pvalue Permutation -Mut_A_Vector- RMA

Fs Pvalue Permutation -Vector_B6- RMA

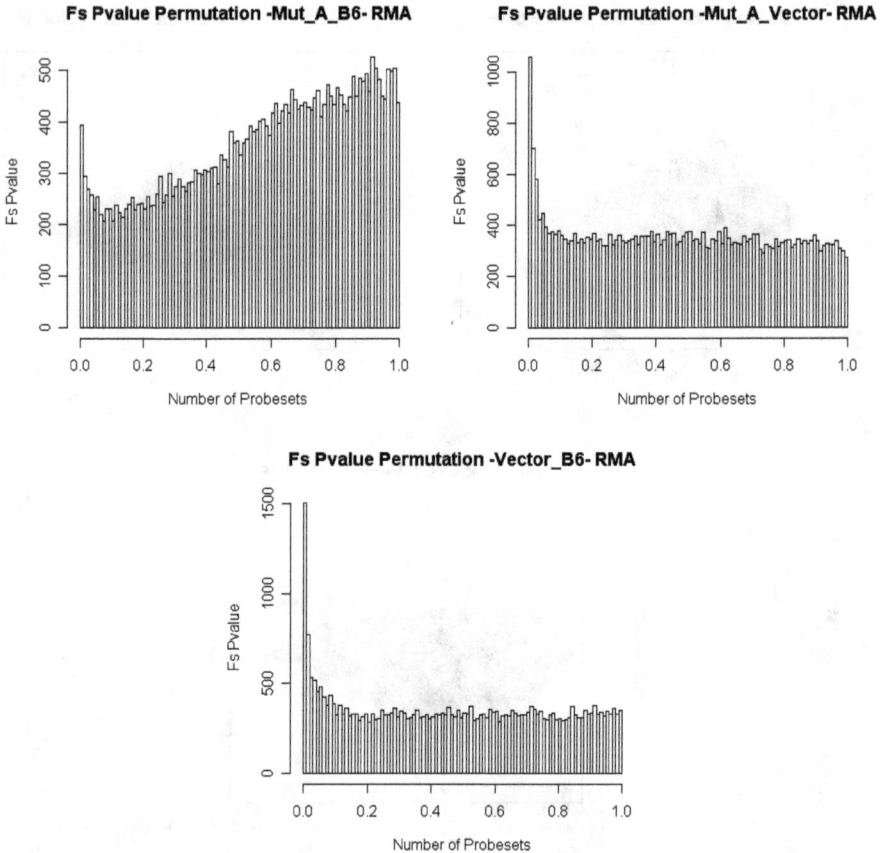

Fig. 3.10.   Histogram of the distribution of Fs permutation *p*-values resulting from the contrast of RMA-processed data for sample groups Mut_A versus B6 (top left), Mut_A versus Vector (top right), and Vector versus B6 (bottom center).

3.3.4. *Scatterplot of normalized intensities*

A scatterplot of $\log_2$-transformed, quantile-normalized intensities allows for the comparison of the degree of similarity (i.e. correlation) of samples within a group to the degree of similarity of samples between groups. Arrays are plotted in groups according to sample type (Figs. 3.6 and 3.7).

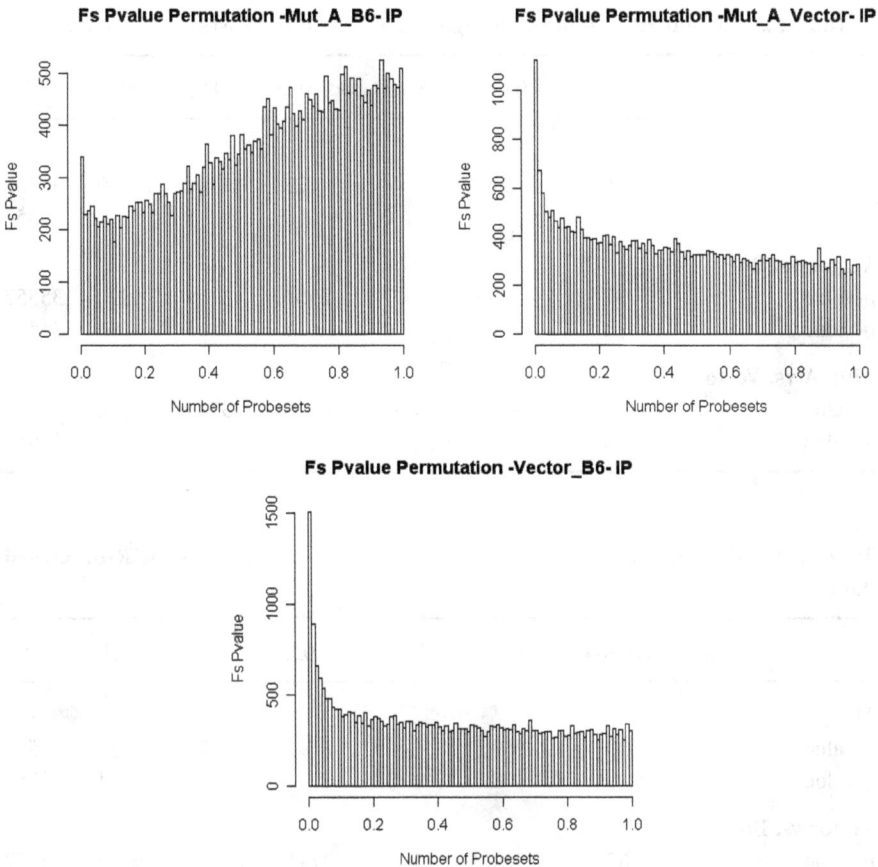

Fig. 3.11. Histogram of the distribution of Fs permutation *p*-values resulting from the contrast of iterPLIER-processed data for sample groups Mut_A versus B6 (top left), Mut_A versus Vector (top right), and Vector versus B6 (bottom center).

3.3.5. Average MA plot of normalized intensities

MA plots are used to visualize the similarity of groups of samples. The *x*-axis represents the average log intensity of one factor level versus another, whereas the *y*-axis represents the log ratio (Figs. 3.8 and 3.9).

Table 3.3.   **Cumulative number of significant calls using RMA-processed data.**

	<1e–04	<0.001	<0.01	<0.025	<0.05	<0.1	<1
Mut_A vs. B6							
p-value	6	56	394	828	1446	2590	35557
q-value	0	0	0	0	**1**	**1**	35557
Vector vs. B6							
p-value	37	238	1507	2556	3789	5890	35557
q-value	0	0	0	1	**5**	**80**	35557
Mut_A vs. Vector							
p-value	32	164	1058	2055	3212	5090	35557
q-value	0	0	0	0	**0**	**33**	35557

Table 3.4.   **Cumulative number of significant calls using iterPLIER-processed data.**

	<1e–04	<0.001	<0.01	<0.025	<0.05	<0.1	<1
Mut_A vs. B6							
p-value	20	67	340	691	1273	2351	35557
q-value	0	0	0	1	**1**	**1**	35557
Vector vs. B6							
p-value	67	293	1507	2741	4197	6445	35557
q-value	0	0	5	28	**84**	**346**	35557
Mut_A vs. Vector							
p-value	45	201	1123	2080	3354	5670	35557
q-value	0	0	0	6	**32**	**110**	35557

3.4. Statistical Analysis

3.4.1. *Analysis of variance (ANOVA) model*

The following ANOVA model was used:

$$y \sim \text{Strain}.$$

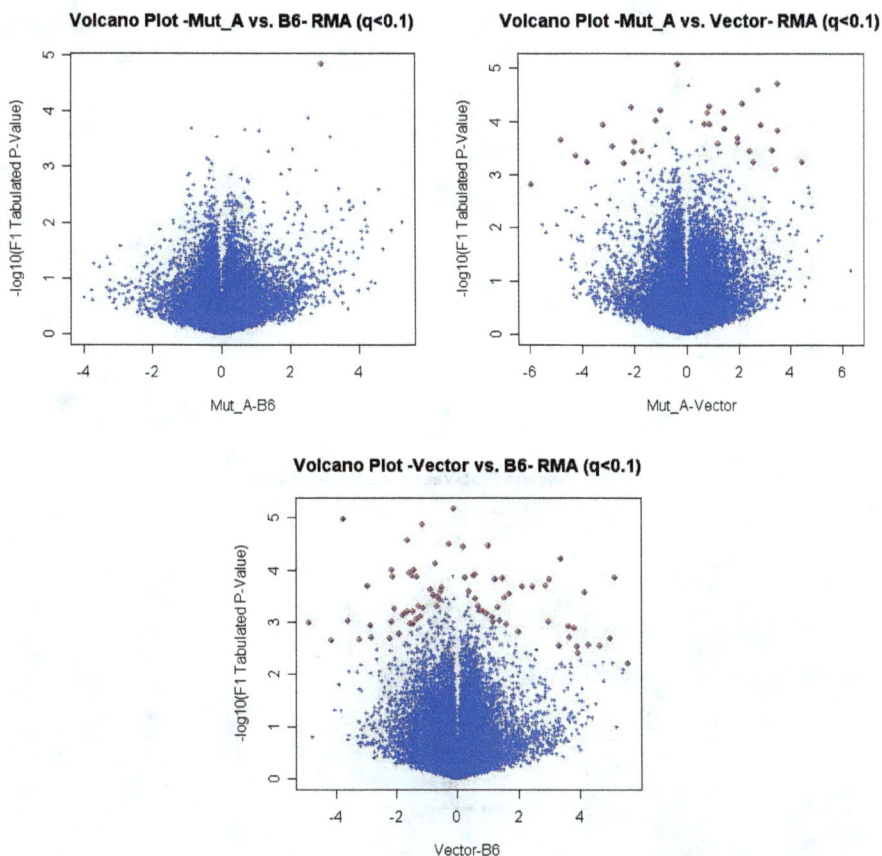

Fig. 3.12. **Volcano plots resulting from the contrasts of RMA-processed data for sample groups Mut_A versus B6 (top left), Mut_A versus Vector (top right), and Vector versus B6 (bottom center).**

3.4.2. *Contrasts*

The following contrasts were used:

Mut_A	versus	B6
Mut_A	versus	Vector
Vector	versus	B6

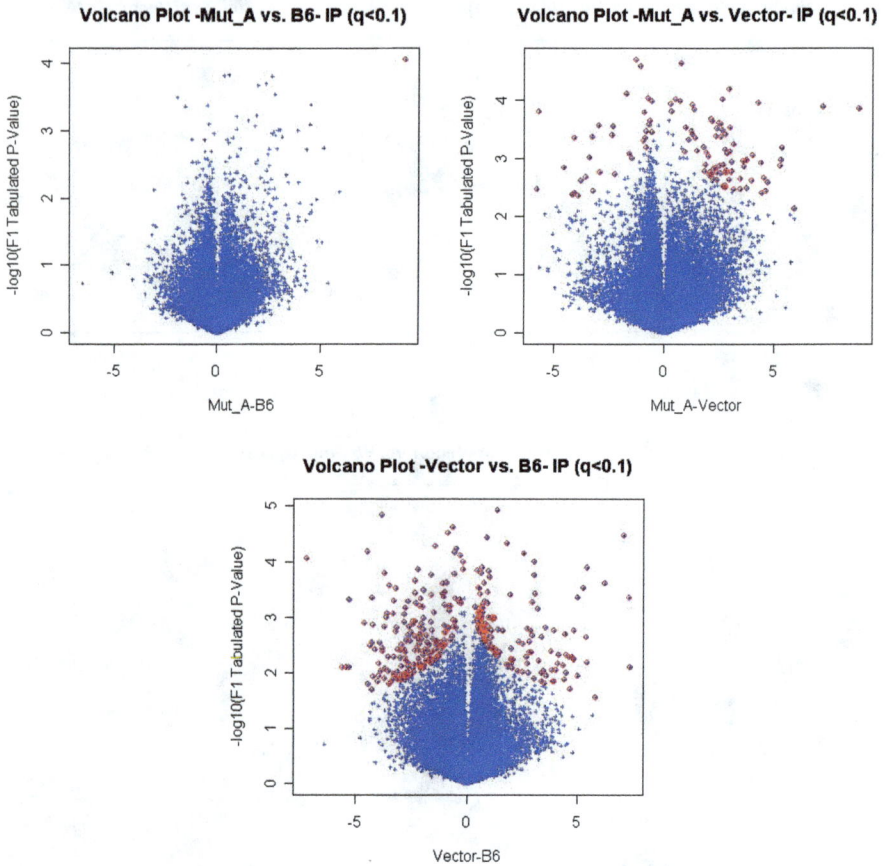

Fig. 3.13.   Volcano plots resulting from the contrasts of iterPLIER-processed data for sample groups Mut_A versus B6 (top left), Mut_A versus Vector (top right), and Vector versus B6 (bottom center).

3.4.2.1. *Fs permutation p-value distribution*

The distribution of Fs permutation p-values is shown in Figs. 3.10 and 3.11.

3.4.2.2. *p- and q-value summary*

RMA summary

Table 3.3 shows the number of differentially expressed probes at various p- and q-value thresholds for the contrast using RMA-processed data.

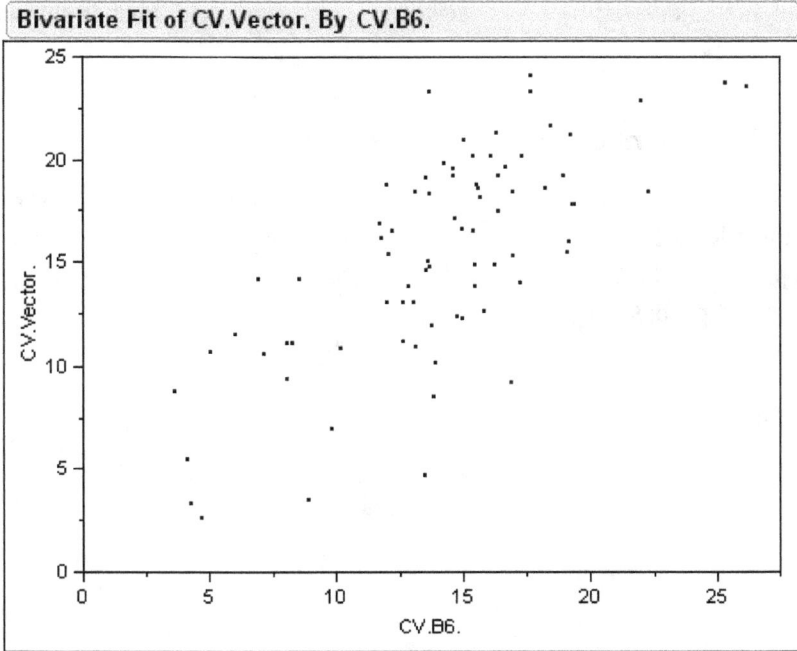

Fig. 3.14.   **Mean coefficient of variation for sample group B6 versus mean coefficient of variation for sample group Vector.**

Differentially expressed genes are declared at an FDR q-value threshold of 0.05. Due to the stringency of FDR, a lower threshold, $q < 0.1$, was selected to identify differentially expressed genes.

iterPLIER summary

Table 3.4 shows the number of differentially expressed probes at various p- and q-value thresholds for the contrast using iterPLIER-processed data. Differentially expressed genes are declared at an FDR q-value threshold of 0.05. Due to the stringency of FDR, a lower threshold, $q < 0.1$, was selected to identify differentially expressed genes.

3.4.2.3. *Volcano plot*

Volcano plots in Figs. 3.12 and 3.13 show the relationship between fold change and the level of significance as represented by the F_1 test statistics, which are calculated by probeset-specific error variance. Probes shown in red have a q-value < 0.1.

3.5. Coefficient of Variation Analysis

Analyzing the spread of mean coefficient of variation data across sample groups allows for the identification of outlier probesets, which may contain probe-level intensity trends that may be indicative of alternative splicing or bad probes (Fig. 3.14).

CHAPTER 4

A Novel Method for DNA Microarray Data Analysis: SDL Global Optimization Method

This chapter is concerned with the challenge of mining knowledge from DNA microarray gene expression data. With the objective to discover unknown patterns from microarray data, methodologies are derived from machine learning, artificial intelligence, and statistics. Nowadays, microarray expression data accumulate at an alarming speed in various storage devices, and so does valuable information. However, it is difficult to understand information hidden in data without the aid of data analysis techniques. Both machine learning and data mining have been applied to the field in order to better understand microarray expression datasets. A data mining system usually enables one to collect, store, access, process, and ultimately describe and visualize datasets. The discussion of data collection and storage is not included here, though it is important for mining microarray expression data.

Data mining has successfully provided solutions for finding information from data in many medical research fields, such as bioinformatics and pharmaceuticals. Many important problems have been addressed by data mining methods, such as neural networks, fuzzy logic, decision trees, genetic algorithms, and statistical methods. Data mining tasks can be descriptive and predictive; in other words, it is an interdisciplinary field with a general goal of predicting outcomes and uncovering relationships in data (Han and Kamber, 2001; Hand *et al.*, 2001; Kantardzic, 2002; Mitra *et al.*, 2002). Microarray data analysis is one of the most attractive fields of data mining. With the help of gene expressions obtained from microarray technology, heterogeneous cancers can be classified into appropriate subtypes (Schena *et al.*, 1995). Many different kinds of machine learning and statistical methods have recently been applied to analyze gene expression data (Alizadeh *et al.*, 2000; Brown *et al.*, 2000; Deutsch, 2003; Khan *et al.*, 2001). However, our experience tells us that there is a need to analyze the expression of all genes detected by DNA microarray technology at the same time, as genes always work in groups but not individually. So far, a method that allows simultaneous analysis of more than 30000 genes is still lacking. Here, we introduce our novel analytical method for microarray data, using our leukemia study as an example. This method is called SDL global optimization.

4.1. Research Subjects

Human Philadelphia chromosome-positive (Ph⁺) leukemia is one of the most commonly occurring blood cancers, and accounts for about 20% of all leukemias. While the new therapeutic drug Gleevec is effective in treating Ph⁺ patients, its effectiveness in some patients is compromised due to the development of clinical drug resistance. Improved therapy will depend on identifying key genes that play significant roles in leukemia development. Using our novel SDL-optimization method for gene classification and other soft computing strategies, one can perform accurate and efficient large-scale analysis of gene expression data to identify genes affected in leukemia cells, and the identified genes will have the potential to serve as diagnostic markers and therapeutic targets for leukemia patients. It will also allow monitoring disease progression during treatment to reduce debilitating side effects. Moreover, strategies learnt from our leukemia research can be applied to patients with other types of cancers. Developing novel techniques for improving early diagnosis using microarray data will be at the cutting edge, and considerable potential technology transfer opportunities may arise from the outcomes.

As stated in Chapter 2, the BCR-ABL oncogene is the cause of Ph⁺ leukemias. The BCR gene, on chromosome 22, breaks at either exon 1, exon 12/13, or exon 19; and fuses to the c-ABL gene on chromosome 9 to form, respectively, three types of the BCR-ABL chimerical gene: P190, P210, or P230. Each of these three forms of the BCR-ABL oncogene is associated with a distinct type of human leukemia. The P190 form is most often present in B-ALL, but only rarely in CML (Deininger *et al.*, 2000); whereas P210 is mainly involved in CML, and in some acute lymphoid (Deininger *et al.*, 2000) and myeloid leukemias in CML blast crisis. P230 is found in a very mild form of CML (Pane *et al.*, 1996). Ph+ B-ALL and lymphoid blast crisis of CML account for about 20% of adult cases and 5% of childhood cases of acute B-lymphoid leukemia. Among those patients with BCR-ABL-induced B-ALL, 50% of adults and 20% of children carry the P210 form of BCR-ABL while the rest of the patients carry the P190 form (Deininger *et al.*, 2000; Druker *et al.*, 2001a; Sawyers, 1999). It is still not fully understood why the three forms of BCR-ABL induce distinct diseases, but it is important to reveal the underlying mechanisms.

This chapter systematically presents how to discover useful information based on the DNA microarray expression data collected from our leukemia studies. Specifically, the chapter will, step by step, describe the analysis of the microarray data obtained from our experiments studying gene expression profiles for P190, P210, and P230 BCR-ABL. In addition, data mining tasks normally include data preprocessing, data modeling, and knowledge description. To cover this area of data mining using our novel SDL global optimization method, we also analyze some publicly available microarray data on leukemia to show the advantages of our data mining method.

4.2. Experimental Design

4.2.1. Rationale

DNA microarrays usually consist of thin glass or nylon substrates containing specific DNA gene samples spotted in an array by a robotic printing device (Anonymous, 2002). Researchers spread fluorescently labeled mRNA from an experimental condition onto the DNA gene samples in the array. This mRNA binds strongly with some DNA gene samples and weakly with others, depending on the inherent double-helical characteristics. A laser scan is done on the array and sensors to detect the fluorescence levels (using red and green dyes), indicating the strength with which the sample expresses each gene. The logarithmic ratio between the two intensities of each dye is used as the gene expression data. The relative abundance of spotted DNA sequences in a pair of DNA or RNA samples is assessed by evaluating the differential hybridization of the two samples to the sequences on the array.

Gene expression levels can be determined for samples taken at multiple time instants of a biological process or under various conditions. Each gene corresponds to a high-dimensional row vector of its expression profile (Mitra and Acharya, 2005). The goal of microarray experiments is to identify genes that are differentially expressed in the conditions being studied (Draghici, 2002). To compare gene expression profiles among P190, P210, and P230, we expressed these three genes, respectively, in a B-lymphoid cell line using a parental cell line that does not express BCR-ABL as a control.

4.3. Fold Change Analysis

Fold change is defined as follows:

Take $\log_2$-transformed normalized intensities from RMA (robust multi-chip averaging) for two samples;

Let $a = \log_2$ (intensity sample #1)
Let $b = \log_2$ (intensity sample #2)
If $a > b$: fold change $= 2\ (|a - b|)$
If $a < b$: fold change $= -2\ (|a - b|)$
If $a = b$: fold change $= 0$.

Preprocessing of the gene expression profile is often necessary to reach the goal of converting from raw data to biological significance. The following steps are common:

- normalizing the hybridization intensities within a single array experiment;
- transforming the data using a nonlinear function, like the logarithm in case of expression ratios;
- estimating and replacing missing values in expressions, or adapting existing algorithms to handle missing values;

- filtering the gene expression profile to eliminate those that do not satisfy some simple criteria; and
- standardizing or rescaling the profiles to generate vectors of length one.

Mouse 430 v2 Affymetrix GeneChip arrays are used for all experiments. Probe intensity data as CEL files are imported into the R software environment (http://www.R-project.org). Probe-level data quality is assessed using image reconstruction, histograms of raw signal intensities, and MA plots. Normalization is performed for each batch separately using the RMA (affy/RMA; http://www.bioconductor.org) method, using all probe intensity data sets together to form one expression measure per probe set per array.

Fold change analysis is conducted for each pair of comparisons as listed above. The results are summarized in two sets of files. The first file lists fold change for every probe set on the microarray; the second file lists fold change of ±2 or more. Once opened in Microsoft Excel and sorted by the column of fold change in descending order, the probes that showed the largest fold changes will present at the top and the bottom of the sorted list.

In these files, the first four column names are as follows (from left to right):

> *Clone id* = name of the gene probes
> $Log2(rma.expr\_P190)$ = $\log_2$-transformed RMA-processed
> expression for P190
> $Log2(rma.expr\_P210)$ = $\log_2$-transformed RMA-processed
> expression for P210
> *Relative.fold.change_P190.relative.to.P210*
> = 2 ^ absolute.value.of(Log2(rma.expr_P190)
> − Log2(rma.expr_P210))
> = expression ratio of P190 vs. P210.

For example, probe 1455238_at has a relative fold change of 19.50; this means that it is upregulated in P190 compared with P210, and that P190 is 19.50 times higher than P210. On the other hand, probe 1450140_a_at has a relative fold change of −7.605; this means that it is downregulated in P190 compared with P210, and that P210 is 7.6 times higher than P190.

Exploratory hierarchical clustering analysis is conducted for classifying samples across time points. A heat map (a color image of expression level with two-way clustering dendrogram added) is also generated to visualize the samples. The filtered gene lists were generated by selecting genes that have the largest variation across samples across both time points. The coefficient of variation (CV), computed as standard deviation divided by the mean, is used as a measure of the magnitude of variation. Three thresholds — CV > 0.1, CV > 0.15, and CV > 0.2 — were used to generate three gene lists for clustering. The distance metric, (1 – correlation), and Ward method were used in the calculation.

Taking the 12-hour time point data as an example, Fig. 4.1 shows a bivariate fit of P190 by P210 with marked *p*-values of 0.9, 0.95, and 0.99.

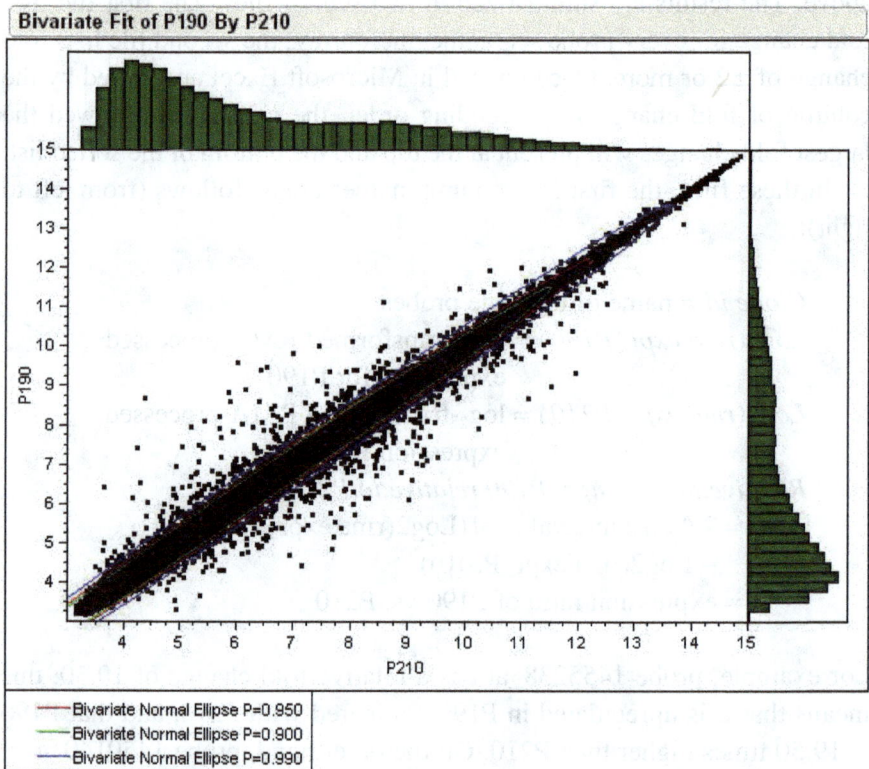

Fig. 4.1.   Bivariate fit of P190 by P210 data.

Array preprocessing and fold-change analysis are conducted for each time point/batch separately. The microarray expression data used involved 45 101 genes. Data distribution at the 12-hour time spot is shown in Fig. 4.2 for P190, P210, and P230 samples, respectively.

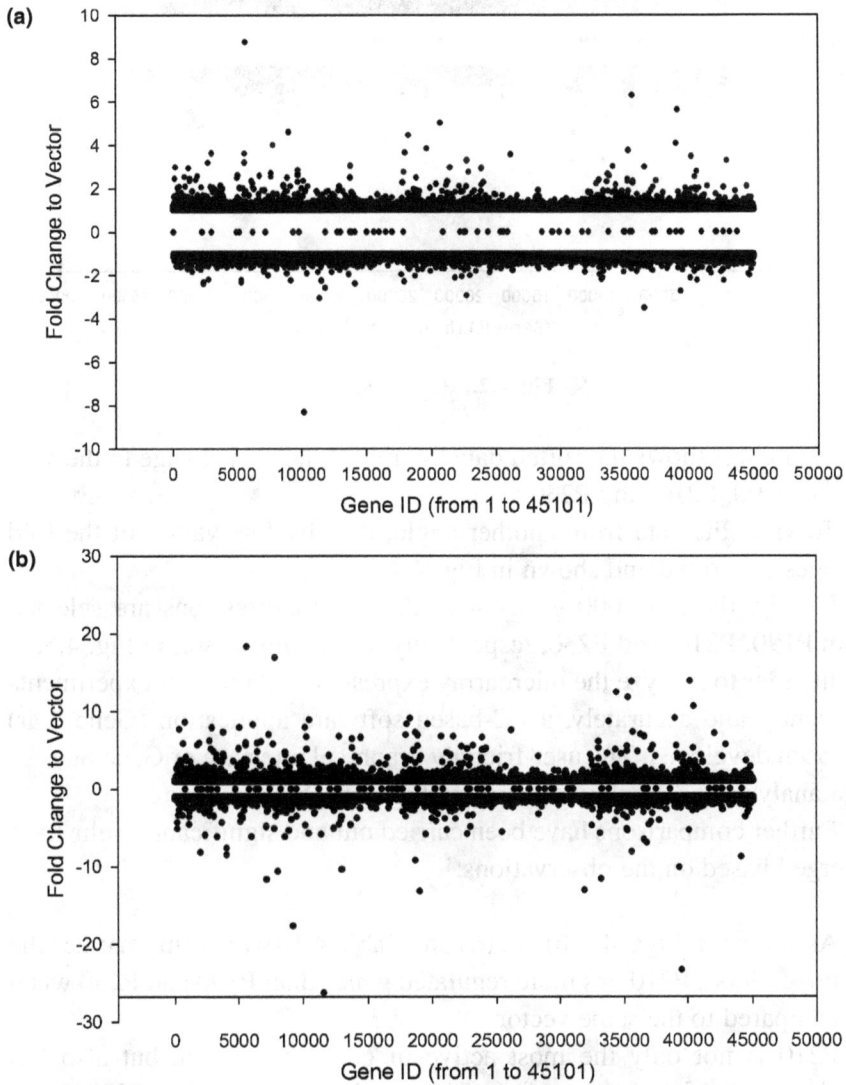

(a)

(b)

Fig. 4.2. (a) P190 raw data. (b) P210 raw data. (c) P230 raw data.

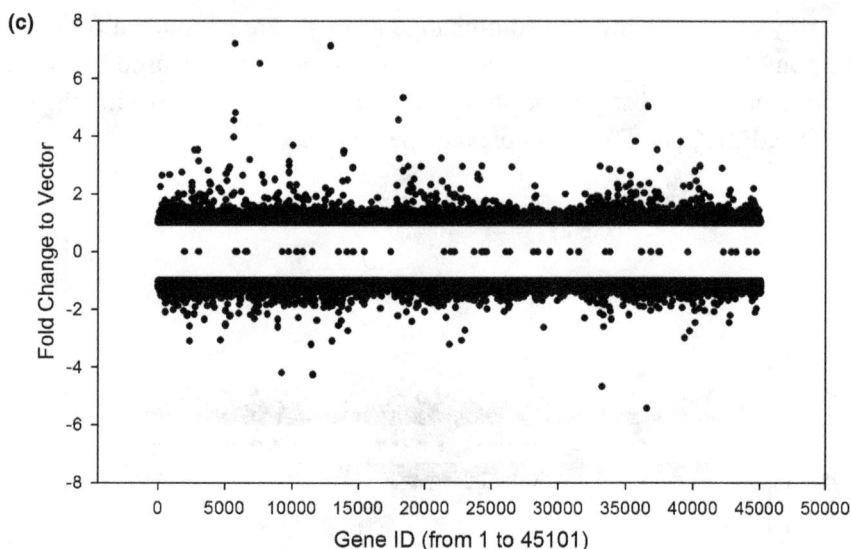

Fig. 4.2. (*Continued*)

Figure 4.3 shows the sorted data to present the fold change to the vector for P190, P210, and P230.

To view the data from another angle, the absolute values of the fold changes are sorted and shown in Fig. 4.4.

Finally, the top 1000 genes with the most expressions are selected from P190, P210, and P230, respectively, for a comparison in Fig. 4.5.

In order to analyze the microarray expression data for our experiments efficiently and accurately, a PC-based software application (Gene Star) has been developed. The user-friendly graphical interface of Gene Star for data analysis and display is shown in Fig. 4.6(a).

Further comparisons have been carried out and significant results have emerged based on the observations:

- As shown in Figs. 4.6(b)–4.6(d) and Table 4.1 (which summarizes the three plots), P210 has more regulated genes than P190 and P230 when compared to the same vector.
- P210 is not only the most active in gene expression, but also has almost all of the same expressed genes in either P190 or P230. P210 behaves as a "father" of P190 and P230.

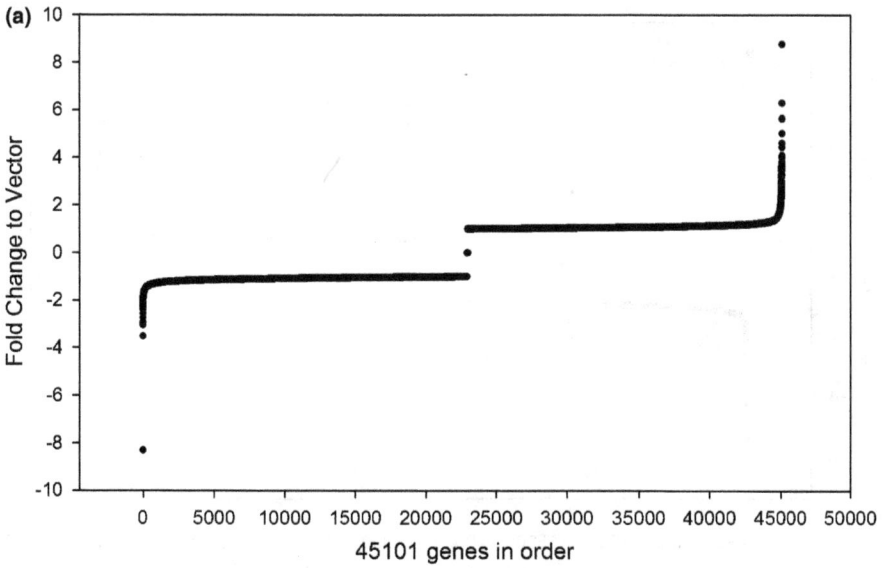

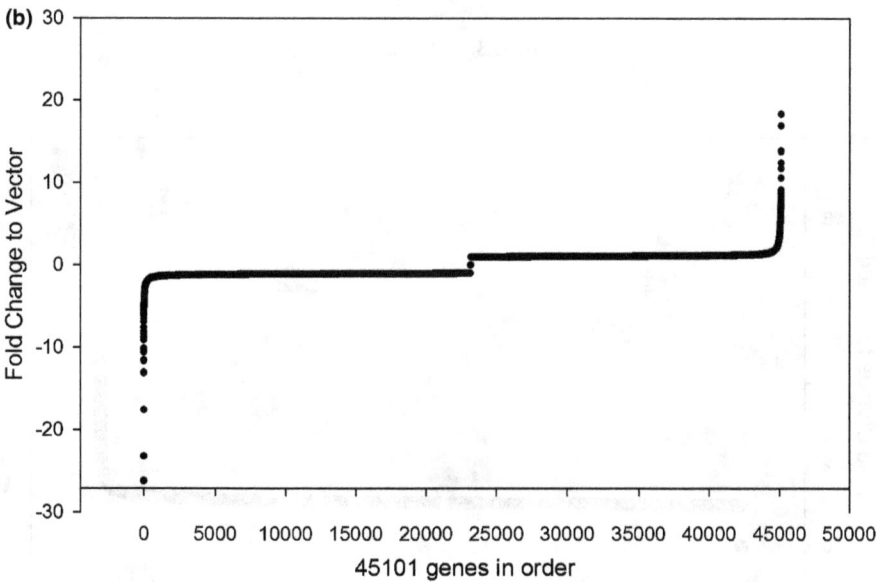

Fig. 4.3. (a) P190 sorted from low to high. (b) P210 sorted from low to high. (c) P230 sorted from low to high.

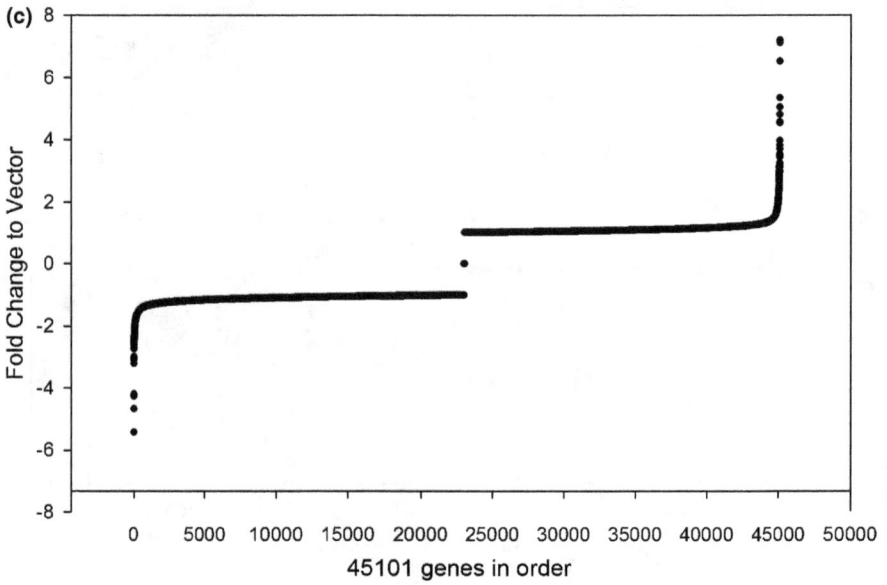

Fig. 4.3.   (*Continued*)

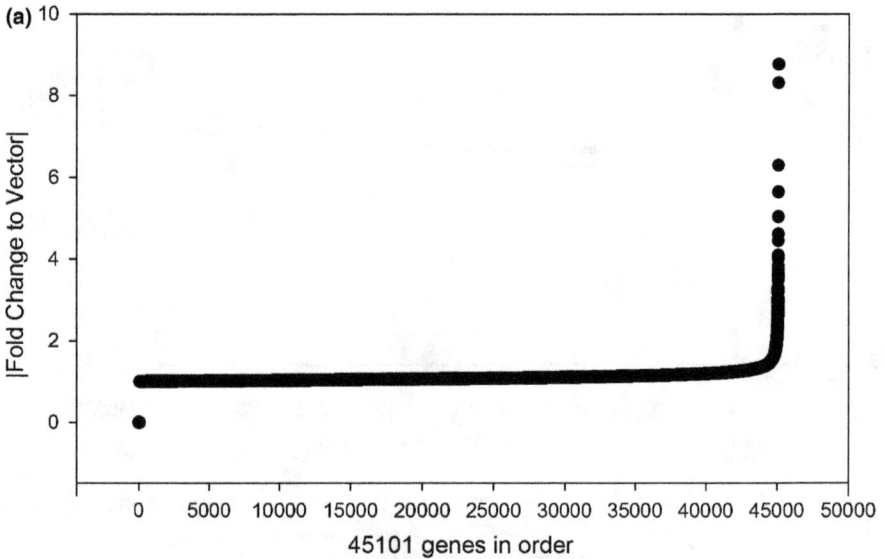

Fig. 4.4.   (**a**) P190 absolute fold change sorted from low to high. (**b**) P210 absolute fold change sorted from low to high. (**c**) P230 absolute fold change sorted from low to high.

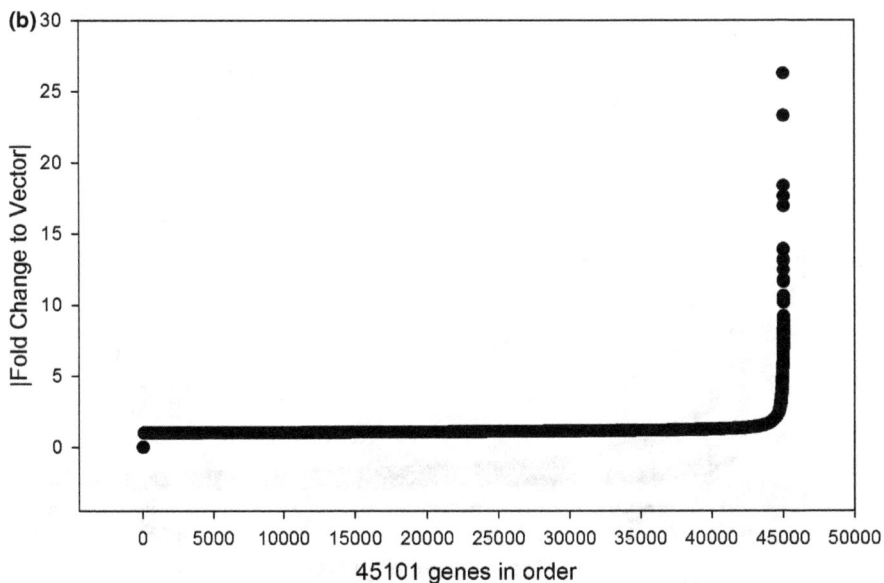

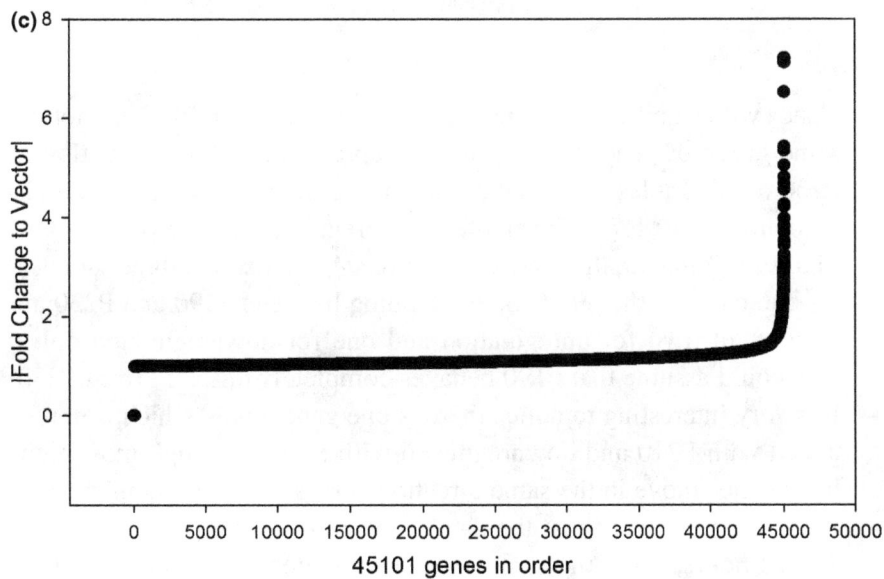

Fig. 4.4. (*Continued*)

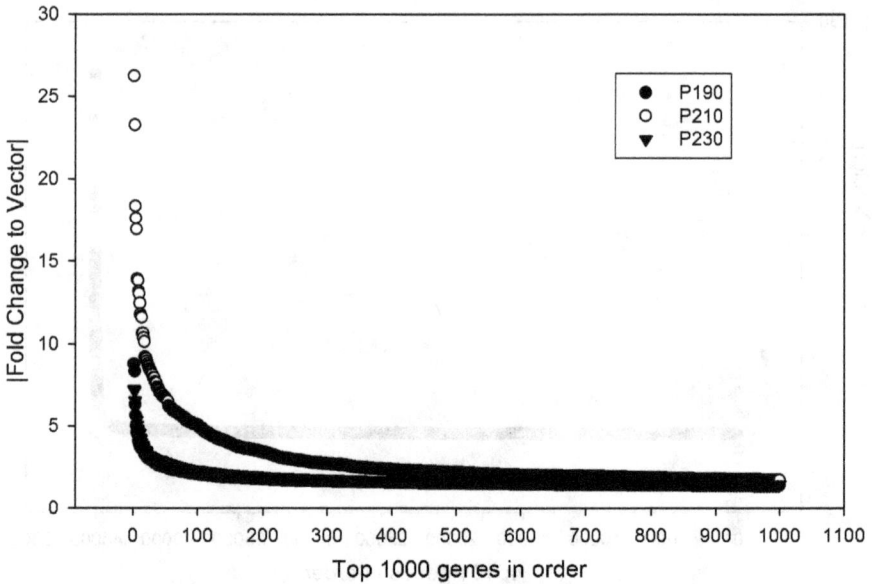

Fig. 4.5.   Top 1000 expression values.

- Genes with significant upregulation/downregulation in P190 are not the same group of genes with significant upregulation/downregulation in P230 at all. Tables 4.2 and 4.3 contain the numbers of the ignorantly regulated genes for each sample and coregulated among the samples. Figures 4.7 and 4.8 illustrate the relationships across the three samples. As one can see, the numbers overlapping between P190 and P230 are very small (two for upregulation and one for downregulation only). One could assume that P190 behaves completely different from P230.
- It is very interesting to notice there is one gene, 4659, which is upregulated with P210 and downregulated with P230, although most regulated genes move in the same direction across the three samples.
- It has been observed that the data arrays at two time points, 12 hours and 18 hours, show the same expression pattern. In other words, the 12-hour data are consistent with the 18-hour data.
- Experiments on the drug effects have also been carried out, and the results will be published elsewhere.

(a)

Top	Gene ID	Clone ID	Chromosome	Value	Fold Change
1	5681	1421375_a_at	3	11.06798656	8.765654575
2	35639	1451344_at	5	10.01089134	6.295436651
3	39133	1454838_s_at	17	6.314859547	5.637478106
4	20754	1436448_a_at	2	8.298212176	5.031820959
5	9106	1424800_at	1	6.153442703	4.605339714
6	18314	1434008_at	NA	9.643296341	4.446730344
7	39006	1454711_at	15	8.056470884	4.085319299
8	7885	1423579_a_at	3	8.231606693	4.010032922
9	19699	1435393_at	NA	7.064296331	3.843356991
10	35328	1451033_a_at	3	9.764220709	3.747949288
11	18051	1433745_at	15	8.681287484	3.650064517
12	3040	1418709_at	7	8.382523573	3.631554522
13	5613	1421307_at	3	8.441771886	3.616699277
14	26193	1441887_x_at	NA	5.21801608	3.566920171
15	40194	1455899_x_at	11	7.235576533	3.476946982
16	22781	1438475_at	NA	9.955465691	3.289343176
17	42982	1458687_at	8	6.711593202	3.281162684
18	5614	1421308_at	3	7.082096606	3.201995128
19	2738	1418407_at	7	7.616702961	3.195117887
20	13876	1429570_at	8	7.662864513	3.04489637
21	34133	1449838_at	17	5.009536308	3.014512252
22	36586	1452291_at	5	7.466451118	2.993062522

Experiment setup — No. of Genes: 45101 — Chromosome: All — Time spot (hours): 12 / 18 — Samples: P190, P210, P230 — Drug: No / Yes — Load Gene Data — Analyse — Close

(b)

Number of Genes with |FC|>2

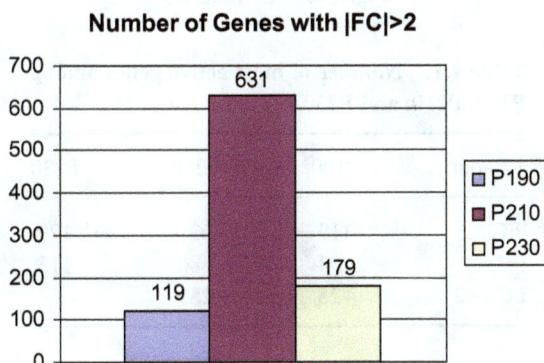

P190: 119 P210: 631 P230: 179

Fig. 4.6. (a) Interface of the software tool. (b) Number of genes with the absolute fold change value >2. (c) Number of genes upregulated with the fold change value >2. (d) Number of genes downregulated with the fold change value <−2.

(c)

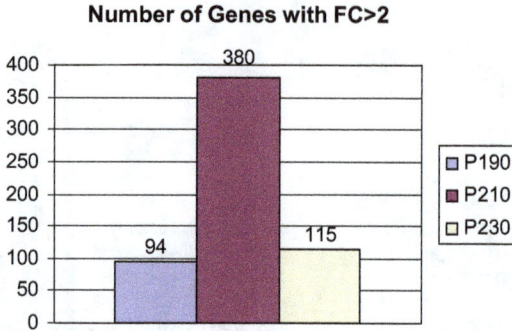

(d)

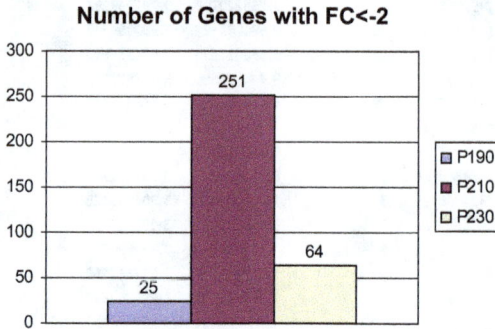

Fig. 4.6. (*Continued*)

Table 4.1. Number of most active genes among P190, P210, and P230.

12-hour	P190	P210	P230
|FC|>2	119	631	179
FC>2	94	380	115
FC<−2	25	251	64

Table 4.2. The number of significantly upregulated genes among P190, P210, and P230 with FC > 2 for P190 and P230 and FC > 3 for P210.

P190 + P210 + P230	P190 + P210	P190 + P230	P210 + P230	P190	P210	P230
54	32	2	45	6	>61	15

Table 4.3. **The number of significant downregulated genes among P190, P210, and P230 with FC < −2 for P190 and P230 and FC < −3 for P210.**

P190 + P210 + P230	P190 + P210	P190 + P230	P210 + P230	P190	P210	P230
19	5	1	27	0	>56	16

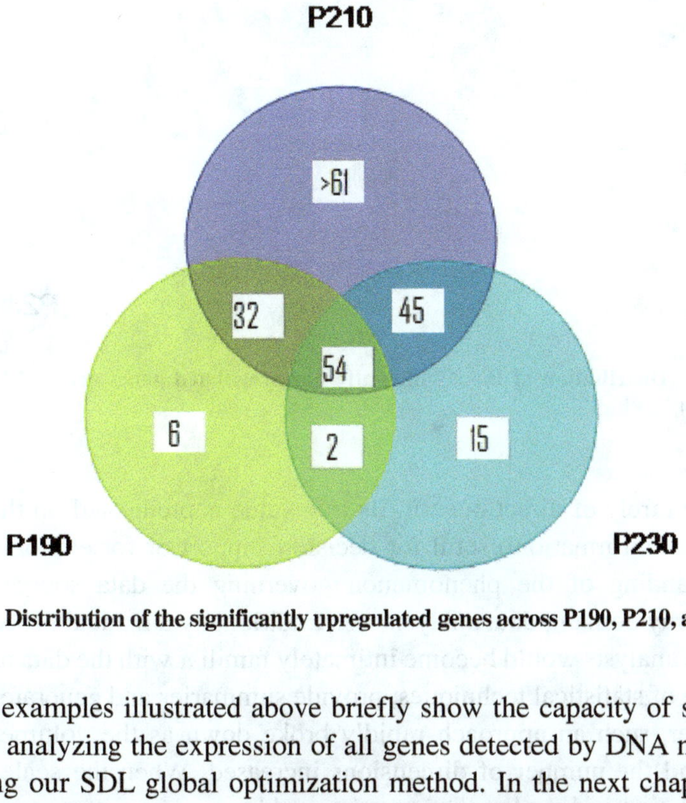

P210

Fig. 4.7. **Distribution of the significantly upregulated genes across P190, P210, and P230.**

The examples illustrated above briefly show the capacity of simultaneously analyzing the expression of all genes detected by DNA microarray using our SDL global optimization method. In the next chapter, we will show more examples of applications of the SDL global optimization method in analysis of our and publicly available microarray data for the molecular classification of leukemia.

4.4. More Information on SDL Global Optimization

Since the fold change does not address the reproducibility of the observed difference and cannot be used to determine the statistical significance, raw

P210

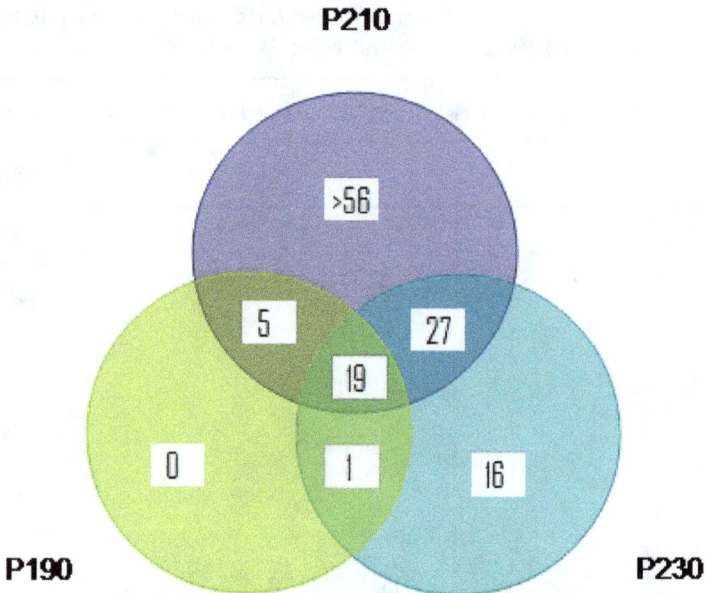

P190   **P230**

Fig. 4.8.   Distribution of the significantly downregulated genes across P190, P210, and P230.

data are rarely of direct benefit. Its true value is predicated on the ability to extract information useful for decision support or for exploration and understanding of the phenomenon governing the data source. In the microarray domain, data analysis was traditionally a manual process. One or more analysts would become intimately familiar with the data and, with the help of statistical techniques, provide summaries and generate reports. However, such an approach rapidly broke down as the volume of data grew and the number of dimensions increased. When the scale of data manipulation and exploration goes beyond human capacities, people need the aid of computing technologies for automating the process. This has therefore prompted the need for intelligent data analysis methodologies, which could discover useful knowledge from data.

Classification is also described as supervised learning (Tou and Gonzalez, 1974). Classification and clustering are two data mining tasks with close relationships. A class is a set of data samples with some similarity or relationship, and all samples in one class are assigned the same class label to distinguish them from samples in other classes. A cluster is a collection of objects that are similar locally. Clusters are usually generated in order to further classify objects into relatively larger and meaningful categories. Clustering is also called unsupervised classification, where no predefined classes are assigned.

According to a data set with class labels, data analysis builds classifiers as predictors for future unknown objects. A classification model is formed first based on available data. Future trends are predicted using the learned model. In the following case, the data sets used are from a public microarray database and the samples are collected to build a model that can be used to classify new samples into categories of ALL or AML for leukemia.

Classification of acute leukemia, having highly similar appearance in gene expression data, has been made by combining a pair of classifiers trained with mutually exclusive features (Cho and Ryu, 2002). Gene expression profiles were constructed from 71 patients having acute lymphoblastic leukemia (ALL) or acute myeloid leukemia (AML), each constituting one sample of the DNA microarray. Each pattern consists of 7129 gene expressions. Feature selection was employed to generate the 25 top-ranked genes for the experiment. A case study from theory to practice is presented in detail in the following sections.

4.4.1. Genetic algorithms (GAs)

GAs are motivated by the natural evolutionary process. Most of the classification techniques with artificial intelligence use GAs as core algorithms. Solutions of the problem at hand are encoded in chromosomes or individuals. An initial population of individuals is generated at random or heuristically. The operators in GAs include selection, crossover, and mutation. To generate a new generation, chromosomes are selected according to their fitness score. The selection operator gives preference to better individuals as parents for the next generation. The crossover operator and the mutation operator are used to generate offspring from the parents. A crossover site is randomly chosen in the parents. The mutation operator is used to prevent premature convergence to local optima (Wang and Fu, 2005). The basic concept in GAs is to introduce effective parallel searching in the high-dimensional problem space.

To solve the problem of mining microarray expression data, GAs are especially useful for the following reasons:

- The problem space is large and complex.
- Prior knowledge is scarce.
- It is difficult to determine a machine learning model to solve the problem due to complexities in constraints and objectives.
- Traditional search methods, such as stochastic, combinatorial, and classical (so-called "hard") optimization-based techniques, perform badly.

4.4.2. *SDL global optimization algorithms*

Although GAs are popular and useful, many problems at hand cannot be resolved easily and accurately. This section combines a powerful algorithm (Li, 2004), which has been used in optical coating design, with the methods of cancer diagnosis through gene selection and microarray analysis. We name this novel analytical method of DNA microarray data "SDL global optimization". A generic approach to cancer classification based on gene expression monitoring by DNA microarrays is proposed and applied to a test leukemia case. By using the orthogonal arrays for sampling and a search space reduction process, a computer program has been written that can operate on a personal laptop computer. The leukemia microarray data can be classified 100% correctly without previous knowledge of their classes.

CHAPTER 5

Applications of the SDL Global Optimization Method in DNA Microarray Data Analysis

5.1. Leukemia Cell Line Study

5.1.1. *Introduction*

In Chapter 2, we introduced our microarray study on a leukemia cell line expressing three forms of BCR-ABL (P190, P210, and P230). The goal of this study is to identify differences in gene expression among the three forms of BCR-ABL. The SDL global optimization method allows us to analyze the expression of all genes simultaneously.

5.1.2. *Datasets*

Here, we analyze two sets of microarray data (P190 and P210), with the fold changes for 21 934 mouse genes.

5.1.3. *Analysis strategies*

5.1.3.1. *Data sorting based on 20 chromosome groups (Fig. 5.1)*

$$\text{Average change } (j) = (\Sigma e(i, j))/n(j).$$

In this equation, $e(i, j)$ are expression fold changes and $n(j)$ are the numbers of genes in every chromosome group; $j = 1, 2, \dots, 20$; $i = 1, 2, \dots, n(j)$.

Average Fold Change based on Chromosome groups

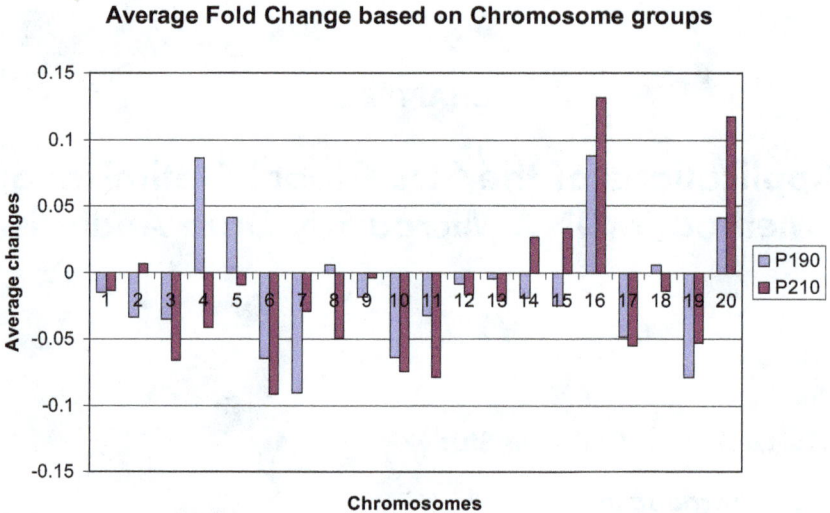

Fig. 5.1.   **Genes altered by P190 or P210 BCR-ABL on each mouse chromosome.**

The No. 4 chromosome shows the strongest changes in two ways. One is that most genes in P190 are upregulated and most genes in P210 are downregulated, while in many other chromosome groups both P190 and P210 move in the same direction. Another finding is that the absolute change of genes on the No. 4 chromosome is the biggest among the 20 chromosomes.

5.1.3.2. *Dynamics of gene expression*

In order to enhance the signal-to-noise ratio (SNR), the dynamics of all gene expression levels were calculated. Figure 5.2 shows that P210 signals are uniformly stronger than the P190 signals, which can be treated as system errors and can be balanced out easily if necessary.

$$\text{Average absolute change } (j) = (\textstyle\sum |e(i, j)|)/n(j).$$

Here, $e(i, j)$ are expression fold changes and $n(j)$ are the numbers of genes in every chromosome group; $j = 1, 2, \ldots, 20$; $i = 1, 2, \ldots, n(j)$.

Dynamics of Gene Expression

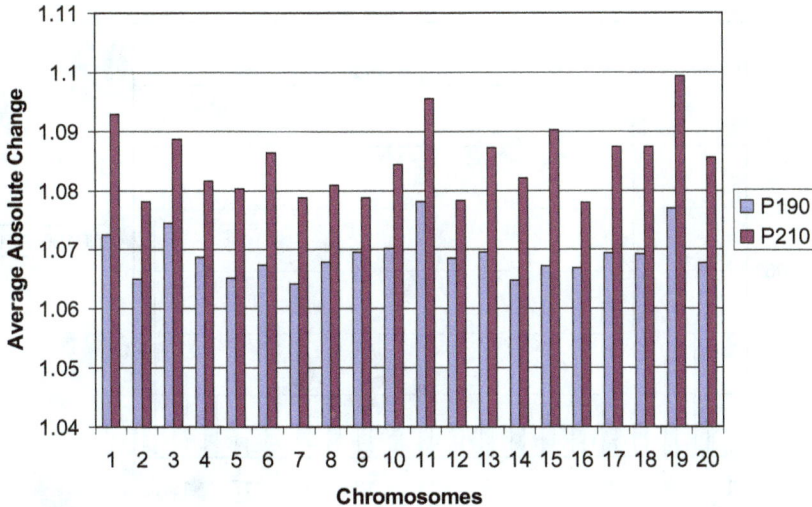

Fig. 5.2.   Average gene expression changes on each chromosome.

5.1.3.3. *Absolute difference between P190 and P210 (Fig. 5.3)*

Absolute difference $(j) = (\sum|e_{p190}(i,j) - e_{p210}(i,j)|)/n(j).$

Here, $e(i,j)$ are expression fold changes and $n(j)$ are the numbers of genes in every chromosome group; $j = 1, 2, \ldots, 20$; $i = 1, 2, \ldots, n(j)$.

We find that the No. 4 chromosome has the highest differential expression across the two datasets.

5.1.3.4. *Prefiltering process (gene selection)*

There were only a small number of genes in the P190 and P210 datasets that were differentially expressed across the two datasets. Genes with lower standard deviations across the two samples were removed. A total of 8832 genes in the P190/P210 data set were selected, the deviation ranging from 2 to 6.68.

Difference between P190 and P210

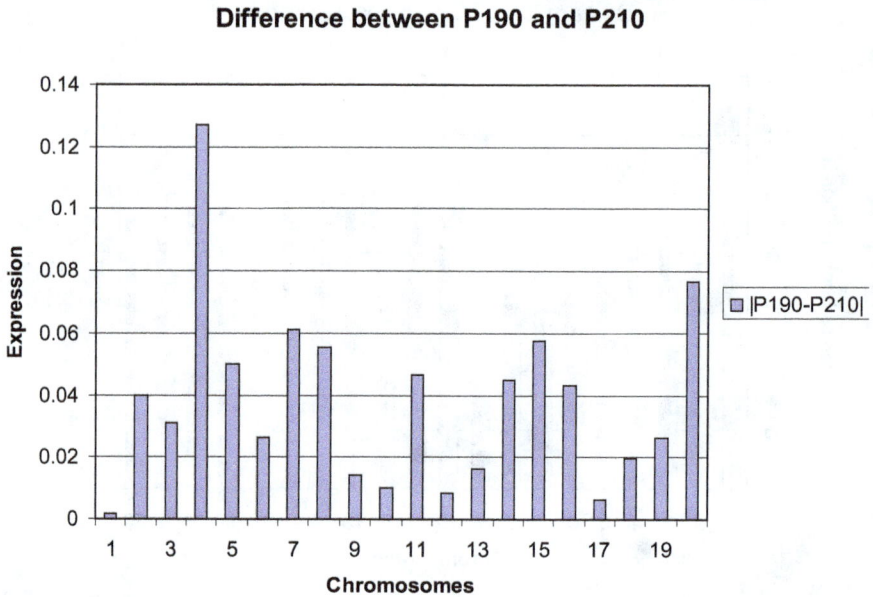

Fig. 5.3.   **Difference in expression levels between P190 and P210 for genes on each chromosome.**

The deviation is defined as follows:

$$D(i) = \left| e_{p190}(i) - e_{p210}(i) \right| \quad i = 1, 2, \ldots, 21934,$$

where $D(i)$ is the deviation of the ith gene, and $e_{p190}(i)$ and $e_{p210}(i)$ are the expression levels of the ith gene for P190 and P210, respectively.

The data were sorted from low to high according to the deviation value across the two samples. The distribution of all the genes is shown in Fig. 5.4. There are two very sharp jumps in the curve, one in the middle and another near the end.

The first jump occurs at number 13 102 (see Fig. 5.5). The deviation changes from about 0 to 2. This fact clearly indicates that about two thirds of the 21 934 genes are not significant at all. Therefore, only 8832 genes are selected for further analysis.

Distribution of Genes

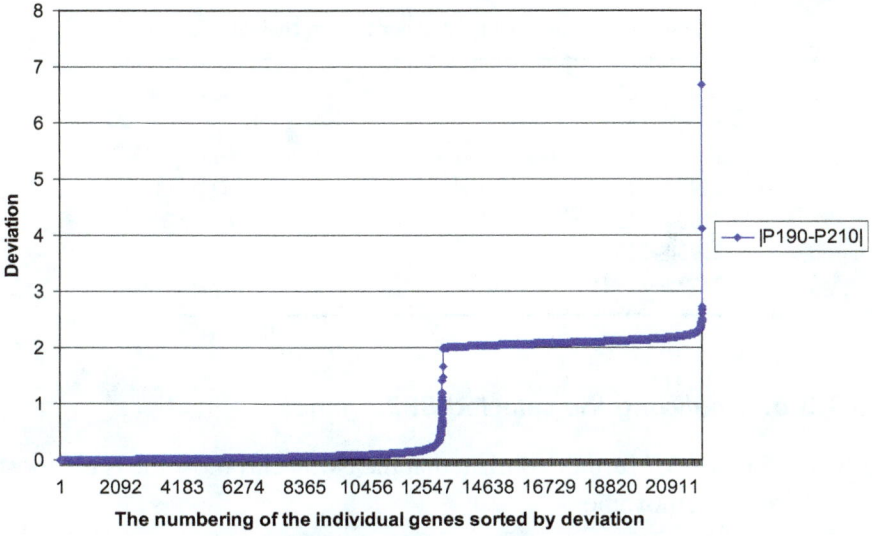

Fig. 5.4.   Gene expression levels for all genes.

Enlarged part of above chart

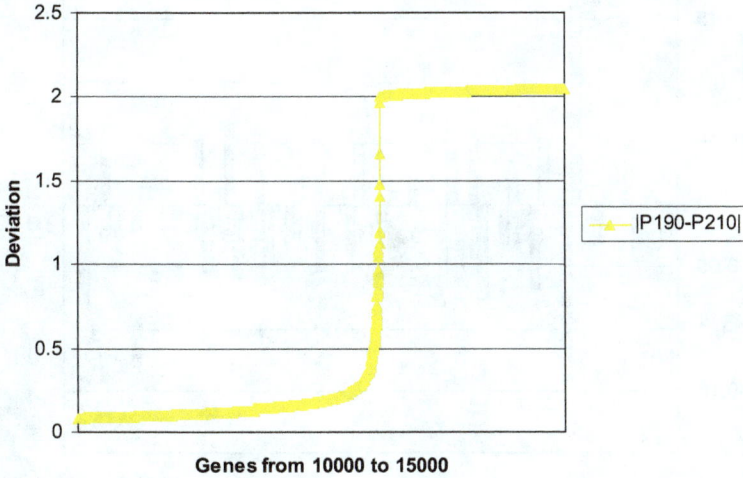

Fig. 5.5.   Detailed analysis of expression levels of 15 000 genes.

5.1.3.5. Grouping genes based on the deviations

Table 5.1. Genes grouped according to their deviation.

Number of genes	Deviation
21934–13102 = 8832	$D \geq 2$
21934–21787 = 147	$D \geq 2.3$
21934–21862 = 71	$D \geq 2.35$
21934–21897 = 37	$D \geq 2.4$
21934–21921 = 13	$D \geq 2.5$

5.1.3.6. Analyzing the selected 8832 genes

Figures 5.6 and 5.7 further confirm that the No. 4 chromosome is the most significant chromosome.

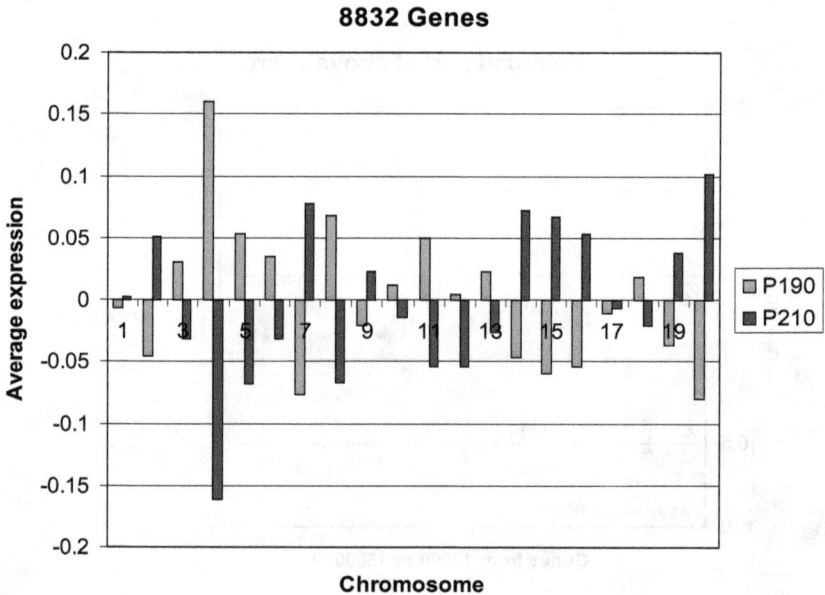

Fig. 5.6. Detailed analysis of expression levels of genes on some chromosomes.

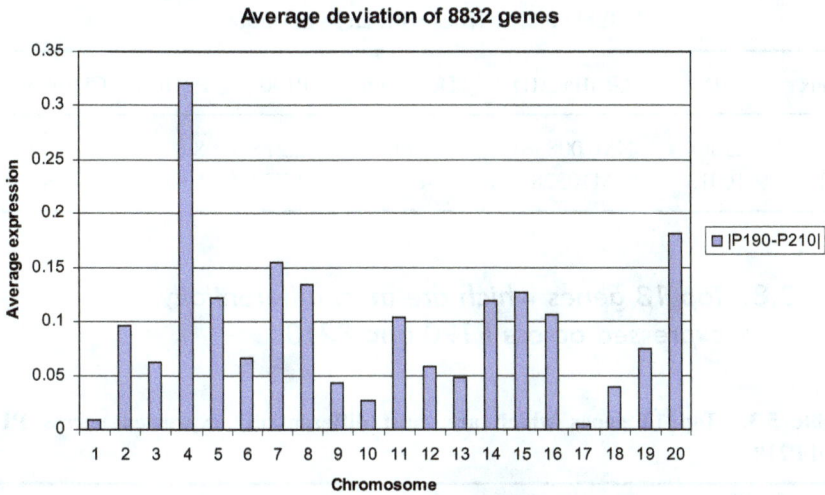

Fig. 5.7. Average deviation of genes altered by BCR-ABL on each chromosome.

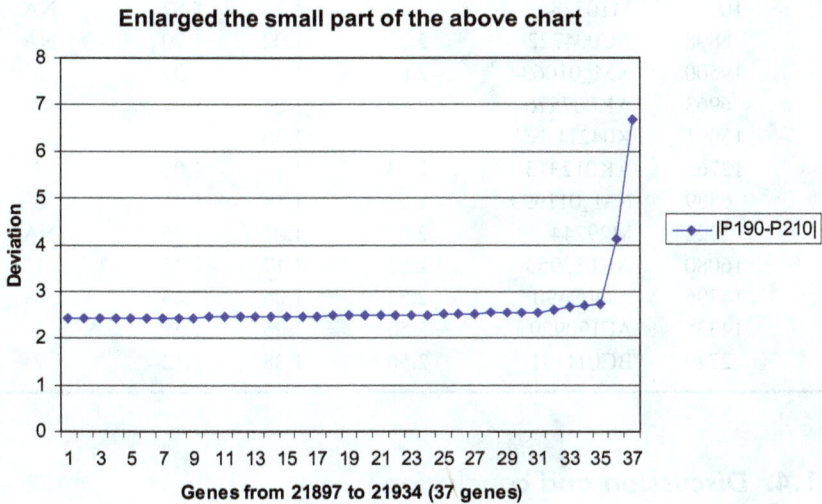

Fig. 5.8. Detailed analysis of deviation for 37 genes.

5.1.3.7. *The second sharp jump of deviations*

Figure 5.8 includes the last 37 genes and shows there are only two genes with huge changes. The latter are given in Table 5.2.

Table 5.2.   Genes with large deviations.

Rank	ID	GenBankID	Deviation	P190	P210	Chromosome
1	21068	NM_008581	6.68	2.21	8.89	NA
2	10013	M10328	4.12	1.75	5.87	NA

5.1.3.8. Top 13 genes which are most differentially expressed across P190 and P210

Table 5.3.   Top 13 genes which are most differentially expressed across P190 and P210.

Rank	ID	GenBankID	Deviation	P190	P210	Chromosome
1	21068	NM_008581	6.68	2.21	8.89	NA
2	10013	M10328	4.12	1.75	5.87	NA
3	8898	BC004722	2.72	1.02	1.70	NA
4	19500	NM_010634	2.69	1.34	1.35	3
5	6963	AK007576	2.66	1.04	1.62	2
6	15091	X04211	2.61	1.36	1.25	NA
7	12765	AK012313	2.53	1.48	1.05	4
8	6539	NM_011999	2.53	1.18	1.35	6
9	5222	M29244	2.53	1.07	1.46	NA
10	16080	AK013953	2.52	1.17	1.35	1
11	14396	AJ409490	2.52	1.06	1.44	NA
12	14335	AF190929	2.50	1.25	1.25	6
13	2749	BC011191	2.50	1.38	1.12	7

5.1.4. Discussion and conclusion

- Not all 21 934 genes are relevant to this experiment; 13 102 could be ignored.
- About one third (8832) of all the genes tested should be looked into.
- Among the 8832 genes, those belonging to the No. 4 chromosome should be investigated closely.

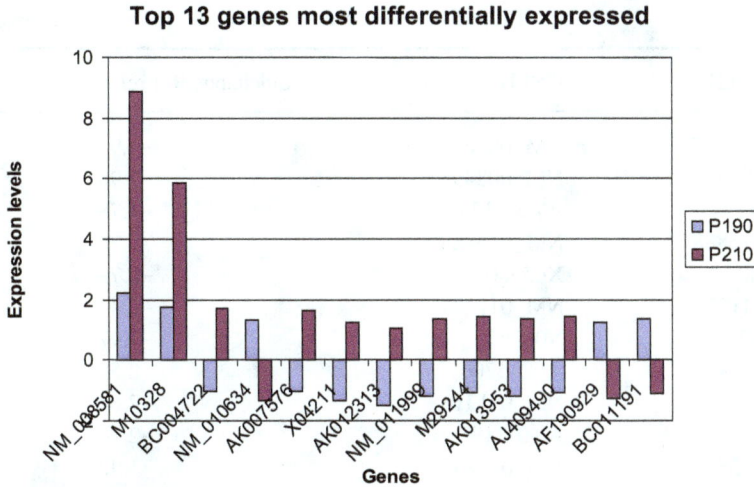

Top 13 genes most differentially expressed

Fig. 5.9.   Analysis of top 13 genes that are most differentially expressed.

- Two abnormally differentially expressed genes (AF190929 and BC011191) need to be explained.
- The top 13 genes most differentially expressed as a group can be used as the experimental targets for further research (Fig. 5.9).
- Comparing the top 13-gene group with the top 29 expressed genes in P190 and the top 22 genes in P210, one can find almost nothing in common (Tables 5.4 and 5.5). This supports the statement that data analysis strategies play a critical role in microarray data interpretation.

5.2. Analyses of Publicly Available Human Microarray Data

5.2.1. Introduction

A generic approach to cancer classification based on gene expression monitoring by DNA microarrays is proposed and applied to two test cancer cases: colon cancer and leukemia. The study attempts to analyze multiple sets of genes simultaneously for an overall global solution to the gene's joint discriminative ability in assigning tumors to known classes. With the workable concepts and methodologies described here, an

Table 5.4. Top hits in P190.

GenBankID	NCBI Hyperlink	Foldchange.Ref.relative.to.Treatment
NM_007645	NM_007645	−1.97
NM_008495	NM_008495	−1.98
AY004174	AY004174	−1.78
NM_016888	NM_016888	2.70
X65980	X65980	−2.79
NM_011224	NM_011224	−2.47
X06762	X06762	−1.97
AF245700	AF245700	2.58
NM_011313	NM_011313	−7.69
AK011193	AK011193	−2.14
NM_023893	NM_023893	1.92
AF217545	AF217545	2.96
M32071	M32071	2.49
AY043479	AY043479	2.47
NM_009061	NM_009061	−2.08
AK017862	AK017862	−1.87
AK007639	AK007639	2.07
AF263458	AF263458	−2.99
NM_013486	NM_013486	−2.09
NM_033264	NM_033264	2.85
NM_011808	NM_011808	1.81
AJ416093	AJ416093	−1.91
K03466	K03466	2.97
X00496	X00496	−2.45
NM_009801	NM_009801	−2.18
NM_021611	NM_021611	−2.87
AK011429	AK011429	−1.90
NM_007763	NM_007763	−1.82
AK003524	AK003524	−1.88

accurate classification of the type and seriousness of cancer can be made. By using the orthogonal arrays for sampling and a search space reduction process, a computer program has been written that can operate on a personal laptop computer. Both the colon cancer and the leukemia microarray data can be classified 100% correctly without previous knowledge of

Table 5.5. Top hits in P210.

GenBankID	NCBI Hyperlink	Foldchange.Ref.relative.to.Treatment
NM_023279	NM_023279	−2.26
S60870	S60870	−2.56
NM_008495	NM_008495	−2.52
X89686	X89686	2.23
NM_011224	NM_011224	−2.45
M92334	M92334	−2.55
NM_016866	NM_016866	2.58
BC011154	BC011154	2.50
NM_011313	NM_011313	−7.97
U26473	U26473	−2.67
AY043479	AY043479	3.30
AF263458	AF263458	−3.90
NM_009638	NM_009638	−2.39
X00496	X00496	−3.65
NM_009801	NM_009801	−3.25
NM_010735	NM_010735	−3.90
L46814	L46814	−2.21
NM_009829	NM_009829	−3.80
NM_021611	NM_021611	−4.88
AK011429	AK011429	−3.09
NM_008581	NM_008581	8.89
X80951	X80951	−2.32

their classes. The classification processes are automated after the gene expression data are inputted. Instead of examining a single gene at a time, the SDL method can find the global optimal solutions and construct a multi-subset pyramidal hierarchy class predictor containing up to 23 gene subsets based on a given microarray gene expression data collection within a period of several hours. Such an automatically derived class predictor makes reliable cancer classification and accurate tumor diagnosis in clinical practice possible.

DNA chip technology enables the study of gene expression on a large scale (Barrett, 2005; Churchill, 2002; Enright *et al.*, 1999; Hacia, 1999). Large-scale gene expressions are used to determine drug targets, identify coregulated genes, and study the response to environmental conditions as well as the effect of a single gene or a group of genes on the entire genome (Abruzzo *et al.*, 2005; Bergmann *et al.*, 2003; Debouck and Goodfellow, 1999; Marcotte *et al.*, 1999). Recent advances in biotechnology allow researchers to measure expression levels for thousands of genes simultaneously, across different conditions, and over a specific time period. Analysis of data produced by such experiments offers potential insight into gene functions and regulatory mechanisms (Abul *et al.*, 2005; Alizadeh *et al.*, 2000; Allison *et al.*, 2006; Bowtell, 1999; Brown and Botstein, 1999; Cheung *et al.*, 1999; Duggan *et al.*, 1999; Lipshutz *et al.*, 1999; Perou *et al.*, 1999).

Computation is required to extract meaningful information from the large amount of data generated by expression profiling (Aittokallio *et al.*, 2003; Bassett *et al.*, 1999; Zhang and Gant, 2004). Most of the algorithms commonly applied to microarray data analysis have been correlation-based approaches named cluster analysis (Alon *et al.*, 1999; Cho *et al.*, 2004). An efficient two-way clustering algorithm was applied to a colon cancer dataset consisting of the expression patterns of different cell types; gene expression in 40 tumor and 22 normal colon tissue samples was analyzed across 2000 genes (Alon *et al.*, 1999). Cluster analysis groups genes involved in microarray data that have similar expression patterns. Those clustered genes are likely to be functionally linked and need to be looked into closely. Although cluster analysis has been widely accepted in analyzing the patterns of gene expression, the methods developed may not be able to fully extract the information from the microarray data corrupted by high-dimensional noise. If the noise from the genes that are irrelevant is not sufficiently reduced, incorrect classification for samples or misleading information on selecting informative genes may result. To select informative genes for sample classification, a neighborhood analysis method was developed to obtain a subset of genes that discriminates between acute lymphoblastic leukemia (ALL) and acute myeloid leukemia (AML) successfully (Golub *et al.*, 1999). In the microarray dataset containing 7129 genes, those genes whose expression levels differ significantly in ALL and AML were identified and subsequently used to predict the class membership (either ALL or AML) of new leukemia cases.

Both approaches described above (Alon *et al.*, 1999; Golub *et al.*, 1999) were focused on comparing samples in each single-gene dimension, and assumed that the relevant genes were similarly and uniformly expressed among samples of each type. A multivariate approach that compares samples in a multi-gene dimension using genetic algorithms (GAs) was proposed (Li *et al.*, 2001a). Samples were classified based on the class membership of their k-nearest neighbors (KNN) in the gene space. The dimensionality (length) of the gene subset was arbitrarily set to 50. GAs were used to select hundreds and thousands of subsets of 50 genes that could potentially discriminate between two classes of samples (tumor and normal tissues). The frequency with which genes were selected was statistically analyzed in the large number of 50-dimension gene subsets. The most frequently selected 50 genes were used to predict 34 new samples. Although the performance of the GA predictor with 50 genes was remarkable, only 29 of 34 test samples were correctly predicted with high confidence (Li *et al.*, 2001b). To improve the success rate of classification, more reliable and accurate algorithms are needed.

Many machine learning and data mining technologies have recently been introduced into the field of microarray data analysis to process many subsets of genes simultaneously (Anderle *et al.*, 2003; Brown *et al.*, 2000; Ooi and Tan, 2003; Wren *et al.*, 2004). It is obvious that there is no one feasible approach to evaluate all possible subsets of genes in a given dataset consisting of several thousands of genes. Even with a moderate number of gene elements in a gene subset and a small number of choices for each gene element, the number of possible gene combinations for the gene subset increases rapidly. The true magnitude of the problem can be seen by considering a scanning approach, which measures the objective function value for every possible combination of genes. For example, let us consider scanning a 10-gene subset using the colon data with 2000 genes (2000 gene expression measurements per sample); the total number of possible combinations is approximately more than 10^{30} (2000! divided by 1990!), which would take years for even a supercomputer to complete. Efficient algorithms are needed to sample from fewer subsets to find the best-performing subsets (optimal or near-optimal solutions). Obviously, the problem is one of optimization or global optimization. In order to solve "hard" problems such as gene selection, classification, and clustering, suitable optimization algorithms must be used.

During the past five decades, the field of global optimization has been growing at a rapid pace and many new theoretical, algorithmic, and computational contributions have resulted (Horst and Pardalos, 1995). Global optimization is concerned with the computation and characterization of global minima (or maxima) of nonlinear functions. Global optimization problems are widespread in the mathematical modeling of real-world systems for a very broad range of applications. The majority of problems can be described as some form of global optimization procedures. In the gene selection problem, one would need to find out how to form gene subsets to obtain the optimum classification response — changing one gene element in a given subset may improve the classification performance of the subset at one testing sample, but worsen it at another.

An objective function is necessary to evaluate how close each gene subset gets to the target requirement. The gene selection process involves finding the gene subset that corresponds to the minimum (or maximum) of the objective function. Plotting the objective function against the gene search space of each gene element in the gene subset, one axis per gene element would be needed, plus the orthogonal axis for the objective function. The objective function plot would appear as a multi-peak, multi-variable plot. Because there is an enormous number of interrelated possible gene combinations, the best gene subset cannot be found by any simple process. It is not obvious how to select the genes analytically to find the best solution. The methods currently used in gene selection — such as clustering, neighborhood analysis, and genetic algorithms (GAs) — almost all depend on a starting condition, either selected by the user or generated internally by the program, that is sometimes not obvious. Changing the initial conditions will give a different result, and one has no way of knowing how much improvement could be effected.

Currently available multi-variable optimization algorithms for selecting the gene subset may not give optimum solutions. Usually, those algorithms obtain their final solutions either from optimizing a starting guess or by techniques which may or may not involve a pseudo-random process that gives different answers every time, depending upon the initial conditions. A true global optimization algorithm should always find the very best solution possible within the boundary conditions stipulated. The possibility of creating a true global optimization algorithm for a large number of interdependent variables has been proposed in this study. Although many optimization algorithms may be appropriate for the gene classification problem, SDL global optimization was proposed and applied to cancer classification in this study for its superb performance in theory and applications.

It is a challenge to discover the optimum gene subset solutions from a microarray gene expression system with a large number of interacting gene variables. It is also well known that orthogonal arrays (OAs) have a number of advantages when they are used in designs of experiments (Dey and Mukerjee, 1999; Hedayat *et al.*, 1999). With the help of an established objective function based on *k*-nearest neighbors (KNN), SDL global optimization combines an OA sampling procedure with some search space reduction strategies for constructing a multi-subset class predictor with a pyramidal hierarchy in order to predict the types of tumor tissues correctly.

With the SDL global optimization algorithms proposed in this research, one knows that the solution gene subsets found are optimized within the criteria set — there is no need to try other starting conditions for the same gene subset structure at a given length, because there are no starting guesses. The algorithms inexorably must find the optimum solution that exists within the boundary conditions. This efficiency has powerful economic consequences. For example, previous solutions needing excessive numbers of genes can now be replaced with fewer genes to get the same classification performance and better confidence. One can improve classification performance as well as offer previously unavailable and undetectable gene subsets as class predictors.

Some strategies of SDL global optimization were first successfully applied to the optical thin film design problem (Li and Nathan, 1996). It was also a candidate for the real function testbed of the First International Contest on Evolutionary Optimization in order to solve ten hard mathematical multivariable optimization problems (Alon *et al.*, 1999; Golub *et al.*, 1999). It is of great interest to develop techniques for extracting useful information from the microarray datasets. In this chapter, we report the application of the SDL global optimization approach for classifying and validating two well-known datasets (Alon *et al.*, 1999; Golub *et al.*, 1999) consisting of the expression patterns of different cell types.

In previous years, many clinicians have been unable to provide a clear-cut classification of cancerous patients, based upon the biopsy. However, with the system proposed here, surveying the expression of thousands of genes is made practical. This chapter outlines a very workable concept which, with more development, will bring groundbreaking new potential for accurate diagnosis. Its biggest advantage lies in the fact that the global optimum is always found with little prior knowledge.

5.2.2. Datasets

Two popular microarray gene expression datasets, for colon cancer and leukemia, were used in this study.

5.2.2.1. Colon data

The original gene expression data were downloaded from the Internet (http://dir.niehs.nih.gov/microarray/datamining/public_html/colon.html. The matrix I2000 contains the expression of the 2000 genes with highest minimal intensity across the 62 tissues (Alon *et al.*, 1999). The genes are placed in order of descending minimal intensity. Each entry in I2000 is a gene intensity derived from the ~20 feature pairs that correspond to the gene on the chip. The data are otherwise unprocessed (for example, it has not been normalized by the mean intensity of each experiment). The "name" file contains the expressed sequence tag (EST) number and description of each of the 2000 genes, in an order that corresponds to the order in I2000. The identity of the 62 tissues is given in the file "tissues data". The numbers correspond to patients, a positive sign to a normal tissue, and a negative sign to a tumor tissue. The data contain the expression levels of 2000 genes across the 62 samples, of which 40 are tumor tissues and 22 are normal tissues. Other researchers indicated that there were five tissue samples (Normal34, Normal36, Tumor30, Tumor33, and Tumor36) identified as likely to have been contaminated (Li *et al.*, 2001a); to avoid having uncertainties, those five samples were removed from the colon cancer dataset. Like the previous study (Li *et al.*, 2001b), the remaining 57 samples were then divided into a training set (the first 40 samples) and a test set (17 samples). The numbers of tumor and normal tissue samples are 27 and 13 in the training set and 10 and 7 in the test set, respectively.

5.2.2.2. Leukemia data

The original data were downloaded from the Internet (http://www.broad. mit.edu/cgi-bin/cancer/datasets.cgi). The data contain the expression levels of 7129 genes across 72 samples, of which 47 are the ALL samples and 25 are the AML samples. These datasets contain measurements corresponding to ALL and AML samples from bone marrow and peripheral blood, which were divided into a training set (38 samples) and a test set (34 samples).

5.3. Overall Methodology

The proposed SDL global optimization method in this study includes the following major steps:

(1) sampling within search spaces by using a suitable orthogonal array instead of conducting a random search;
(2) constructing an objective function for optimization algorithms;
(3) using search space reduction strategies;
(4) searching for global optimal solutions;
(5) building up a multi-subset pyramidal hierarchy class predictor for classification; and
(6) predicting through a voting mechanism.

5.3.1. Orthogonal arrays (OAs) and sampling procedure

OAs were discovered and introduced in the middle of the last century (Rao, 1946; Rao, 1947; Rao, 1949). Many statistical texts on experimental designs include OAs (Cochran and Cox, 1957; Montgomery, 1997). OAs are often employed in industrial experiments to study the effect of several control factors. An OA is a type of experiment where the columns for the independent variables are "orthogonal" to one another.

An OA is a matrix of n rows and k columns, with every element being one of the q levels, and is normally represented in the form of $\mathbf{L}_n(k^q)$. The rows of the OA represent the experiments or tests to be performed. The columns of the OA correspond to the different variables whose effects are being analyzed. The entries in the OA specify the levels at which the variables are to be applied. A typical OA $\mathbf{L}_{12}(11^2)$ is shown below.

1	1	1	1	1	1	1	1	1	1	1
2	2	2	1	2	2	1	2	1	1	1
1	2	2	2	1	2	2	1	2	1	1
1	1	2	2	2	1	2	2	1	2	1
1	1	1	2	2	2	1	2	2	1	2
2	1	1	1	2	2	2	1	2	2	1
1	2	1	1	1	2	2	2	1	2	2
2	1	2	1	1	1	2	2	2	1	2
2	2	1	2	1	1	1	2	2	2	1
1	2	2	1	2	1	1	1	2	2	2
2	1	2	2	1	2	1	1	1	2	2
2	2	1	2	2	1	2	1	1	1	2

Pick any two columns, say, the first and the last columns, from the above table.

1	1
2	1
1	1
1	1
1	2
2	1
1	2
2	2
2	1
1	2
2	2
2	2

Each of the four possible rows — {(1, 1), (1, 2), (2, 1), (2, 2)} — can be seen here, and they all appear the same number of times (three times here); this is the property that makes it an OA. Since only 1's and 2's appear, this is called a two-level array. There are 11 columns, which means that one can vary the levels of up to 11 different variables; and 12 rows, which means that 12 different combinations of variables can be tested in experiments. The aim is to investigate not only the effects of the individual variables on the outcome, but also how the variables interact.

Owen (1992 and 1994) and Loh (1996) describe some uses for randomized OAs in numerical integration, computer experiments, and visualization of functions. These references contain further references to the literature, which in turn provide further explanations.

The OA used in this research is $L_{242}(11^{23})$, which is too large to be shown here. The OA $L_{242}(11^{23})$ has 242 rows (observations or tests), 23 columns (factors or variables), and 11 levels for each factor. The complete $L_{242}(11^{23})$ is available at http://www.scis.ecu.edu.au/dli/.

The OA $L_{242}(11^{23})$ was initially used in selecting a gene subset with 23 gene elements. The search space of 2000 genes in the colon data was divided into 11 levels equally. If all of the genes are assigned a unique ID number from 1 to 2000 and the initial search space ranges from 1 to 2000, then the selected gene IDs are 1, 200, 400, 600, 800, 1000, 1200, 1400, 1600, 1800, and 2000, respectively. As the first row of $L_{242}(11^{23})$ reads (1, 10, 2, 3, 8, 8, 2, 4, 8, 9, 5, 4, 10, 5, 7, 1, 5, 5, 8, 1, 10, 11, 2), the constructed gene subset will read (1, 1800, 200, 400, 1400, 1400, 200, 600, 1400, 1600, 800, 600, 1800, 800, 1200, 1, 800, 800, 1400, 1, 1800, 2000, 200). Since the duplicated gene IDs are not allowed in a gene subset, those repeated gene IDs are shifted forward or backward a little bit. The modified 23-gene subset now reads (1, 1800, 200, 400, 1400, 1399, 199, 600, 1401, 1600, 800, 599, 1799, 799, 1199, 2, 801, 798, 1401, 3, 1798, 2000, 201).

According to $L_{242}(11^{23})$, 242 different 23-gene subsets were created and evaluated with the defined objective function. All 242 subsets were ranked based on their values of objective function. The top 10% performers in classifying the training set were kept, and those gene IDs included in the top 10% gene subsets were ranked in order to work out the minimum ID and the maximum ID. The new, reduced search space ranged from the minimum ID to the maximum ID. The above process was repeated until the search space was small enough (e.g. fewer than 11 genes left) or the objective function could not be improved any further. The rank No. 1 gene subset in the last round of optimization was chosen as the optimal solution for the 23-gene subset. The optimization was run 23 times with different lengths (23, 22, ..., 2, 1) of gene subsets at each run; a total of 23 optimal solutions were obtained. All 23 optimal solutions constructed a multi-subset cancer class predictor and were then used to classify the samples in the test dataset. All 23 gene subsets were arranged to form a pyramidal layer-by-layer hierarchy, with the shortest subset (one gene) at the top and the longest subset (23 genes) at the bottom (see Table 5.7 and Table 5.9 for details).

5.3.2. Objective function

An objective function is also called a fitness or merit function, which is a measure of the ability for a selected gene subset to classify the training set samples according to the SDL optimization procedure. There are several ways, such as neighborhood analysis (Golub *et al.*, 1999), support vector machines (Peng *et al.*, 2003; Liu *et al.*, 2005), and *k*-nearest neighbors (KNN) (Li *et al.*, 2001a), to construct an objective function for the optimization and gene selection algorithms. Among them, KNN is used for the proposed SDL global optimization because it is easy to compute. The Euclidean distance between a single sample (represented by its pattern vector V_m) and each of the pattern vectors of the training set containing M samples is calculated:

$V_m = (g_1, g_2, ..., g_n)$, where n is the number of genes in the vector that can be set from 1 to 23 in order to form the gene vectors (or subsets) with different lengths; g_n is the expression level of the nth gene in the mth sample; $m = 1, 2, ..., M$. For the colon cancer dataset, $M = 40$; for the leukemia dataset, $M = 38$.

Each sample is classified according to the class membership of its KNN as determined by the Euclidean distance in n-dimensional space. If all or a majority of the KNN of a sample belong to the same class, the sample is classified as that class; otherwise, the sample is considered

Table 5.6. KNN rules ($k = 5$).

	Among the ranked 5 nearest neighbors	Classification	Class code
Colon	All 5 are normal samples	Normal	1
	All 5 are tumor samples	Tumor	−1
	4 are normal and 1 is tumor	Normal	1
	4 are tumor and 1 is normal	Tumor	−1
	3 are normal and 2 are tumor	Unknown	0
	3 are tumor and 2 are normal	Unknown	0
Leukemia	All 5 are ALL samples	ALL	1
	All 5 are AML samples	AML	−1
	4 are ALL and 1 is AML	ALL	1
	4 are AML and 1 is ALL	AML	−1
	3 are ALL and 2 are AML	Unknown	0
	3 are AML and 2 are ALL	Unknown	0

unclassifiable. The k was arbitrarily set to 5 in this study. The detailed rules are shown in Table 5.6.

If the class membership of a training set sample and its five nearest neighbors in the particular n-dimensional space defined by a gene subset agree, or four out of five nearest neighbors agree, the sample is classified and a score of 1 is assigned to that sample. These agreement scores are summed across the training set. For convenience, this sum is divided by the number of training samples (40 for colon cancer and 38 for leukemia) as the value of the objective function for the selected gene subset. The bigger the value, the better the selected gene subset performs in classification. The maximal objective function value is 1, which means that all samples in the training set are classified correctly by the gene subset under testing. The goal of the optimization procedure is to discover the optimal gene subset (optimal solution) with the maximal value of the objective function. As in other methods, an objective function is calculated for each subset of genes by the sum over all classifying scores of the samples in the training data set. The optimization process then conducts a search for the gene subset that has the best objective function value (minima or maxima). Therefore, by finding the lowest or highest value of the objective function, one will have discovered the best-performing gene subset. This procedure can be made more sophisticated by introducing weighting factors to increase the importance of user-specified samples in training sets, as well as using other forms of the distance formula between one subset and another.

5.3.3. *Search space reduction for global search*

With local optimization, a fast method for a large number of genes, the program finds the nearest minimum and stops. For some so-called global optimization procedures, the algorithm not only finds a local minimum, but can also find some neighboring minima. The processes, however, is a hit-and-miss situation because starting at a different place can result in different solutions.

The global algorithm in SDL repeatedly narrows the region where the global minimum is known to lie by using a special OA sampling that operates simultaneously in all orthogonal dimensions (one for each gene in the gene subset) to find the optimum solution. As the process runs, one can observe the range of genes for each gene variable in an n-dimensional subset being reduced.

The SDL global optimization algorithm operates to discover the optimum solution. An analogy illustrates the principles involved: assume plotting the objective function against 2000 genes in the colon cancer data with a goal of finding a gene or a gene subset corresponding to the maximum objective function value F (or $1/F$ for the minimum value, for convenience in the illustration). See Fig. 5.10 for a one-dimensional (1D) analogy showing local and global optimization processes.

As discussed earlier, a single objective function number can be used to describe the classification performance of a current gene subset. By plotting a multi-dimensional graph with objective function as one of the axes, one can visualize the process. One requires as many orthogonal axes as the number of variables (genes) plus one for the objective function. Thus, for a two-gene problem, a three-dimensional (3D) plot is required. To see the process used in a simplified form, imagine a two-dimensional (2D) array along the x- and y-axes, which corresponds

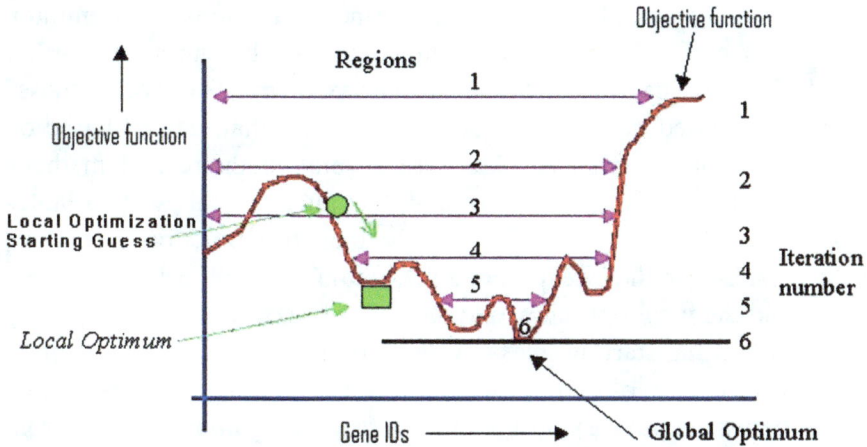

Fig. 5.10. Local optimization methods locate the minimum closest to the starting point. Global optimization techniques may find other local minima, but cannot ensure that the absolute lowest has been found. The region investigated by SDL sampling is progressively narrowed (designated as 1 to 6) to ascertain a true global minimum.

to a two-gene problem. Let the values along the x-axis represent gene IDs of the first gene variable; and values along the y-axis, gene IDs of the second gene variable. The objective function value is plotted in the z-direction.

The task of the process then is to find the x and y values that generate the highest value of the objective function. In this description, one shall invert the objective function back since a peak is easier to see than a valley. One form of objective function can be transformed into the other by taking a reciprocal of it. Therefore, one chooses the form of objective function whose value increases as the design performance improves, i.e. we want to maximize the objective function. The game is to change each gene variable in order to maximize the objective function. To help understand the optimization process, consider the analogy of a man wandering in a cratered terrain with a global positioning satellite (GPS) receiver, which displays his absolute x-, y-, and z-coordinates. Height (z) is the objective function, and x,y represent the gene IDs of each gene variable in a two-gene problem. His task is to find the highest point (largest objective function value) bounded by the user-defined maximum and minimum values of x and y. If he just walks upwards until he can go no further uphill, he will have found the local maximum. The "best guess" approach is based on a starting design (position) that may be based on many years of experience. There are several mathematical methods available, such as the gradient method, that alter several gene variables simultaneously, see what happens to the objective function, and move the gene subset in the direction of a maximum of the objective function. These find the local optimum, and the end subset solution is completely dependent on the starting guess. A derivation of this is the GA method which, by use of evolution in populations and random numbers, is able to find better maxima, exploring other subsets by "jumping away" from the nearest peak.

The GA optimization methods use a completely different technique to optimize the gene subsets. While it has the advantage of being capable of producing complex gene subsets with minimum user interaction, the solution found is unlikely to be the global optimum. The final solution is still dependent on the initial starting gene selection, and many more genes than are necessary are often required for a given performance.

SDL optimization operates by a process of searching for all regions in the gene space where a height greater than a specific level is located. This is akin to creating a contour map by slicing parameter space at a constant value of objective function. The levels (choices) of the genes were equally spaced within every reduced search space.

In Fig. 5.11(a), one sees a representation of a two-gene-subset problem. Using a search analogy, to find the peak, the region in which the highest peak must lie is narrowed each time the plane is raised. This occurs until there is only one peak left. Its coordinates correspond to the gene IDs of the optimum solution. This is the equivalent of a plane parallel to the x-y plane at height z [see Fig. 5.11(b)]. This plane intersects the topography and identifies the entire region within which the peak is known to lie. By raising the slicing plane repeatedly, the region within which the peak must lie is made smaller and smaller until only the highest peak remains [see Fig. 5.11(c)]. Its coordinates correspond to the gene IDs of the optimum gene subset. In practice, the surface is a mathematical

(a)

(b)

(c)

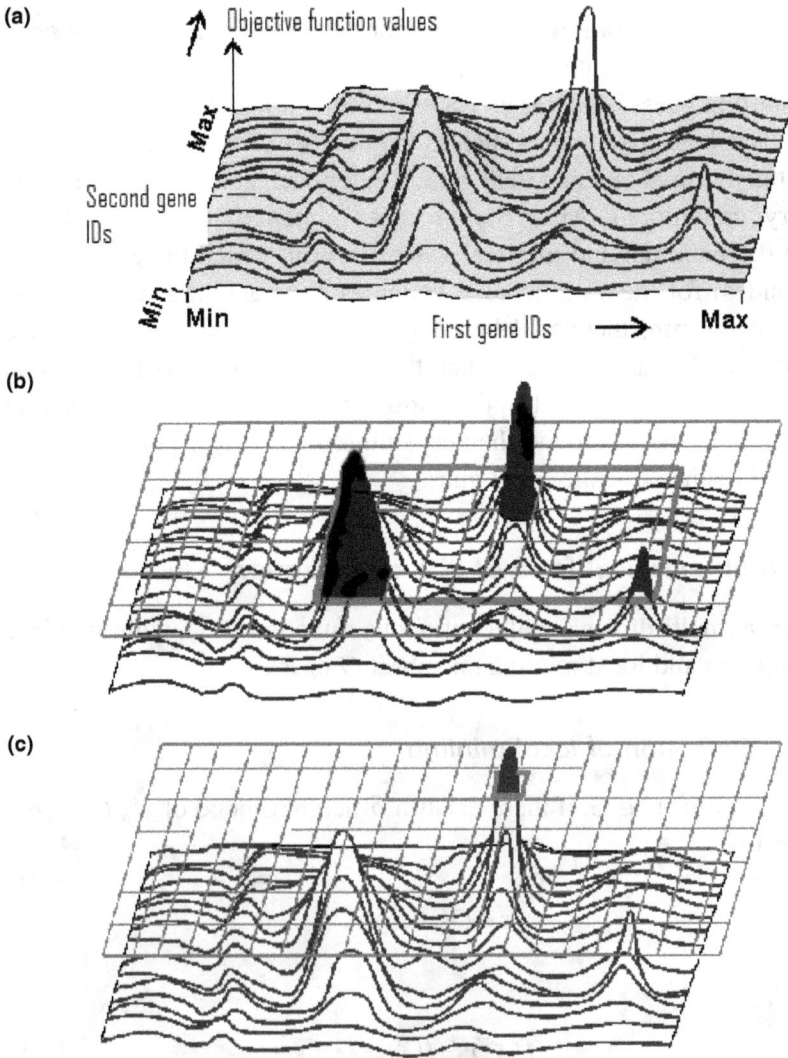

Fig. 5.11. (a) In this description of a 2-gene-subset problem, the boundary conditions are defined by the gene variable's maximum and minimum IDs. One seeks to find the gene IDs that give rise to the maximum value of the objective function. (b) A plane is constructed of "constant objective function", and the boundary of regions having a higher objective function than that of the plane is identified. (c) As the plane is raised, the region within which the peak of the objective function exists is narrowed. The process is repeated until the two gene IDs (coordinates of the peak) which give rise to the highest peak are uniquely identified.

construct of as many orthogonal dimensions as there are genes in the gene subset with a given length.

No starting guess is necessary (or even possible), and the operator only has to define basic parameters such as the number of genes in a gene subset and the minimal and maximal gene IDs for each gene variable (i.e. the boundary conditions). After the process is started, the operator can observe the maximal and minimal values (within which the global optimum resides) for the various gene variables approaching each other. At the end of the run, there will be no gene variable whose search space (maximal to minimal) is greater than the specified value (as little as one gene). By using SDL global optimization strategies, an operator can be assured that he/she has found the best solution physically possible, independent of his/her so-called best guess.

5.3.4. *Mathematical form of SDL optimization*

Consider a multi-dimensional continuous function $f(x)$ with multiple global minima and local minima on subset G of R^n.

5.3.4.1. *Definition of local minima*

For a given point $x^* \in G$, if there exists a δ-neighborhood of x^*, $O(x^*, \delta)$, such that for

$$x \in O(x^*, \delta),$$

and

$$f(x^*) \leq f(x), \tag{1}$$

then x^* is called a local minimal point of $f(x)$.

5.3.4.2. *Definition of global minima*

If for every $x \in G$ the inequality (1) is correct, then x^* is called a global minimum of $f(x)$ on G, and the global minima of $f(x)$ on G form a global minimum set.

5.3.4.3. *How to find the global minima*

Now, for a given constant C_0 such that the level set $H_0 = \{x \mid f(x) < C_0, x \in G\}$ is nonempty, if $\mu(H_0) = 0$, where μ is the Lebesque measure of H_0, then C_0 is the minimum of $f(x)$ and H_0 is the global minimum set.

Otherwise, assume that $\mu(H_0) > 0$ and C_1 is the mean value of $f(x)$ on H_0. Then,

$$C_1 = \frac{1}{\mu}(H_0) \int_{H_0} f(x)d\mu \qquad (2)$$

and

$$C_0 \geq C_1 \geq f(x^*). \qquad (3)$$

One then gradually constructs the level set H_k and mean value C_{k+1} of $f(x)$ on H_k as follows:

$$H_k = \{|f(x) < C_k, \ x \in G|\} \qquad (4)$$

and

$$C_{k+1} = \frac{1}{\mu}(H_k) \int_{H_k} f(x)d\mu. \qquad (5)$$

With the assistance of OA sampling, a decreasing sequence of mean values $\{C_k\}$ and a sequence of level sets $\{H_k\}$ are obtained.

Let

$$\lim_{k \to \infty} C_k = C^* \qquad (6)$$

and

$$\lim_{k \to \infty} H_k = H^*. \qquad (7)$$

It can thus be proven that C^* is the minimum of $f(x)$ on G, and H^* is the global minimum set.

There are several strategies to avoid missing the global optimum when seeking the minimum solution. Among these, the most important step is to select or design a suitable OA with which the function within domains can be repeatedly sampled. The algorithm is automatically constrained to stay within the function domain and will not request function evaluations outside this domain.

There are two stopping criteria possible: either when the target objective function value is reached, or when the maximum domain length is smaller than the user-selected value. In this research, one uses the latter stop criteria, corresponding to the variation possible for each gene element in the subset — which can be as little as one gene. This means that the global minimum has been found for a particular gene selection range of each gene element, with a variation of less than one gene for each gene element. Strictly speaking, then, the global optimum is not defined at a point, but as lying within a region.

5.3.5. *Multi-subset class predictor*

Although SDL optimization will result in an optimal gene subset with a given length, the classification performance varies. It seems that for both the colon cancer and leukemia datasets, there is no guarantee of naming a single gene subset that is capable of classifying all of the samples in the testing set correctly. It is observed that gene subsets with different lengths tend to misclassify or unclassify the different samples in the test datasets. In other words, gene subsets with the same length will always misclassify a few same samples in the test datasets, although those are all the optimal subsets identified by optimization procedures. This indicates that the key factor to improve the signal-to-noise ratio in classifying very noisy data, such as microarray gene expressions, is the length of the gene subset. Based on the above observation, a multi-subset class predictor was constructed for classification by using all 23 optimal gene subsets with the lengths from 1 to 23 genes. The maximal number of genes involved in the predictor is 276 in total. As some of the genes may appear more than one time, the actual number of the unique gene IDs is a bit less and varies from case to case.

5.3.6. Validation (predicting through a voting mechanism)

The established multi-subset class predictor is validated with the testing datasets for both the colon cancer and the leukemia data. Each gene subset in the predictor predicts the class of every sample in the testing datasets independently according to the same KNN rules ($k = 5$) used in the training stage. The predicted class code (in colon cancer data: 1 for normal, -1 for tumor, 0 for unknown; and in leukemia data: 1 for ALL, -1 for AML, and 0 for unknown) is assigned to the particular sample accordingly. Each single class code is treated as a single vote. For each sample in the testing datasets, up to 23 votes contributed by 23 gene subsets in the predictor can be obtained. The final class predicted by the predictor depends on the sign of the sum of the 23 votes of the sample under test. A positive sign indicates that there are more gene subsets in the predictor vote for class 1 (normal for colon cancer and ALL for leukemia), and the sample is finally classified as 1 by the multi-subset predictor. A negative sign indicates that there are more gene subsets in the predictor vote for class -1 (tumor for colon cancer and AML for leukemia), and the sample is finally classified as -1 by the multi-subset predictor. When the sum is 0, there are equal numbers of gene subsets among the 23 gene subsets for class 1 and class -1; in this case, the corresponding sample should be classified as 0 (unknown or unclassified). It is not difficult to interpret the actual values of the classification results. The absolute value of the sum of the 23 votes should indicate the predicting strength. The larger the value is, the more confident the prediction is.

5.4. Experimental Results

A Microsoft Windows-based computer program with a user-friendly graphical interface has been written. The entire experimental computation was carried out on a personal laptop computer (1.7 GHz Intel Pentium Pro/II/III). The software can be downloaded from http://www.scis.ecu. edu.au/dli/ (Li, 2006) and is available free to researchers. Both the colon cancer and the leukemia samples were classified 100% correctly. The classification processes are automated after the gene expression data are inputted. It can find the global optimum solutions and construct a multi-subset class predictor containing up to 23 gene subsets based on a given microarray gene expression data collection, such as the colon cancer or leukemia data, within a period of several hours.

For the convenience of computation, every gene was assigned a unique integer ID number (from 1 to 2000 for colon, and from 1 to 7129 for leukemia) according to the order in their original datasets. The aim was to study how changes in the choices of various gene element variables for a gene subset with a given length affect a response variable (success rate in classifying training samples). For each of the gene elements that are used to form a gene subset, 11 choices (levels) were selected for inclusion in the OA sampling based on $L_{242}(11^{23})$. Those 11 choices of gene IDs were generated by the formula (the length of the current search space divided by 10) at an equal distance. Some shifting on the selected gene IDs was necessary to avoid having any repeating genes in a single gene subset. A total of 242 subsets were evaluated with the objective function in the current iteration, and the 10% top-performing subsets were used to reduce the search space. Only two top-performing gene subsets were passed to the next iteration.

Within the search space of 2000 genes for the colon cancer data and 7129 genes for the leukemia data, SDL global optimization found 23 optimal gene subsets with different lengths, from 1 gene to 23 genes, that formed two pyramidal hierarchy class predictors, respectively (see Tables 5.7 and 5.9). Those gene subsets were assumed to be the best-performing gene combinations for classifying the gene datasets used in this study. The selected gene subsets were then used to classify the test samples in both the colon and leukemia datasets. Tables 5.8 and 5.10 show the classification results. Once the validation of all 23 optimal gene subsets was completed, the proposed multi-subset voting mechanism was adopted. One of the classification results (1 for class 1, −1 for class 2, and 0 for unclassified) was obtained by balancing the votes from the 23 gene subsets for the particular testing sample of interest. It is a process of counting votes to make a final decision on the class of the sample under test. For example, sample N28 in the colon dataset (shown in Table 5.8) receives 23 votes in total from the 23 gene subsets in the class predictor. Among the 23 votes, there are 11 votes of normal, 6 of tumor, and 6 of unknown. The class code (1 for normal, −1 for tumor, and 0 for unknown)

Table 5.7. Optimal gene subsets and selected genes for the colon cancer data class predictor.

Gene subsets	Gene IDs																						
1-gene subset	1227																						
2-gene subset	619	1286																					
3-gene subset	249	550	1102																				
4-gene subset	802	164	693	765																			
5-gene subset	1842	1558	1423	853	581																		
6-gene subset	1774	567	249	1095	164	1539																	
7-gene subset	1042	1286	996	1423	1550	1676	581																
8-gene subset	164	1327	1487	1110	1362	1461	567	249															
9-gene subset	251	1360	206	249	1020	1415	1437	2000	1812														
10-gene subset	1670	773	1611	698	1915	855	1562	1087	245	924													
11-gene subset	581	1735	698	276	118	1247	249	629	1551	504	802												
12-gene subset	1043	2000	662	1351	782	567	164	642	66	245	760	977											
13-gene subset	247	1680	1380	1706	1042	765	1282	1058	1880	633	1208	1196	1674										
14-gene subset	572	1686	1905	1699	1294	1372	897	323	1009	493	1809	295	1216	513									
15-gene subset	1033	1402	1085	1201	679	245	1272	1312	897	66	1020	1436	1066	1623	164								
16-gene subset	853	542	2000	1600	1263	355	1393	1714	898	249	164	911	316	1589	902								
17-gene subset	1424	1644	757	249	1626	1067	804	778	1231	1670	1656	1412	2000	1819	1445	1173	888						
18-gene subset	1802	1171	1138	1663	1657	249	1086	327	594	138	1596	1800	1582	309	1264	1538	228	871					
19-gene subset	741	765	1662	1863	623	394	1217	417	605	2000	1818	1071	567	885	164	665	1943	1366	1908				
20-gene subset	254	980	723	1043	547	1678	1819	1041	1519	813	245	1600	1840	479	288	1873	510	883	1805	1166			
21-gene subset	249	948	904	401	1846	583	1165	264	1539	457	2000	1025	1600	228	1945	1213	1610	344	1360	1430	1850		
22-gene subset	1497	1786	1800	831	1849	239	994	513	1370	857	240	362	1608	1890	228	1761	508	341	586	249	539	340	
23-gene subset	712	1209	422	66	471	1792	1970	164	599	990	1362	1128	1472	1469	1557	1672	245	1665	1103	1070	541	767	1490

Table 5.8.   Validation of the colon cancer data with the multi-subset class predictor.

Gene subsets	Prediction of 17 test samples																	Results of classification			Success rate %
	T28	N28	N29	T29	T31	T32	N32	N33	T34	T35	N35	T37	T38	T39	T40	N39	N40	Correct	Incorrect	Unknown	
1-gene subset	-1	-1	-1	-1	-1	-1	-1	-1	-1	-1	-1	-1	-1	-1	-1	1	-1	11	6	0	64.7
2-gene subset	-1	0	-1	-1	-1	-1	0	-1	-1	-1	0	-1	-1	-1	-1	0	-1	10	3	4	58.8
3-gene subset	-1	1	1	-1	-1	-1	1	1	-1	-1	1	-1	-1	-1	-1	1	1	17	0	0	100
4-gene subset	-1	-1	-1	-1	-1	-1	-1	-1	-1	-1	-1	-1	-1	-1	-1	-1	-1	10	7	0	58.8
5-gene subset	-1	-1	1	-1	-1	-1	1	1	-1	-1	1	-1	-1	-1	-1	0	1	15	1	1	88.2
6-gene subset	-1	0	1	-1	-1	-1	1	1	-1	-1	1	-1	-1	-1	-1	1	1	16	0	1	94.1
7-gene subset	-1	0	1	-1	-1	-1	1	1	-1	-1	1	-1	-1	-1	-1	0	1	15	0	2	88.2
8-gene subset	-1	0	1	-1	-1	-1	1	1	-1	-1	1	-1	-1	-1	-1	1	1	16	0	1	94.1
9-gene subset	-1	-1	1	-1	-1	-1	1	1	-1	-1	1	-1	-1	-1	-1	-1	1	15	2	0	88.2
10-gene subset	-1	1	1	-1	-1	-1	1	1	-1	-1	0	-1	0	-1	-1	-1	-1	13	2	2	76.5
11-gene subset	-1	1	1	-1	-1	-1	1	1	-1	-1	1	-1	-1	-1	-1	1	1	17	0	0	100
12-gene subset	-1	1	1	-1	-1	-1	1	0	-1	-1	1	-1	-1	-1	-1	-1	1	15	1	1	88.2
13-gene subset	-1	-1	1	-1	-1	-1	1	1	-1	-1	0	-1	-1	-1	1	-1	1	13	3	1	76.5
14-gene subset	-1	0	1	-1	-1	-1	1	1	-1	-1	1	-1	-1	-1	-1	0	-1	14	1	2	82.3
15-gene subset	-1	1	1	-1	-1	-1	1	1	-1	-1	1	-1	-1	-1	-1	0	0	15	0	2	88.2

(Continued)

Table 5.8. (*Continued*)

Gene subsets	Prediction of 17 test samples																	Results of classification			
	T28	N28	N29	T29	T31	T32	N32	N33	T34	T35	N35	T37	T38	T39	N39	T40	N40	Correct	Incorrect	Unknown	Success rate %
16-gene subset	−1	0	1	−1	−1	−1	1	1	−1	−1	1	−1	−1	−1	1	−1	1	16	0	1	94.1
17-gene subset	−1	1	1	−1	−1	−1	1	1	−1	−1	1	−1	−1	−1	1	−1	1	17	0	0	100
18-gene subset	0	1	1	−1	−1	−1	1	1	−1	−1	1	−1	−1	−1	1	−1	1	16	0	1	94.1
19-gene subset	−1	−1	0	−1	−1	−1	−1	1	−1	−1	−1	−1	−1	−1	−1	−1	−1	10	6	1	58.8
20-gene subset	−1	1	0	−1	−1	−1	1	1	−1	−1	1	−1	−1	−1	0	−1	0	13	2	2	76.5
21-gene subset	0	1	1	−1	−1	−1	1	1	−1	−1	1	−1	−1	−1	1	−1	1	15	0	2	88.2
22-gene subset	−1	1	0	−1	−1	−1	1	1	−1	−1	1	−1	−1	−1	−1	−1	1	16	1	0	94.1
23-gene subset	−1	1	0	−1	−1	−1	0	0	−1	−1	1	−1	−1	−1	−1	−1	1	14	1	2	82.3
Sum of votes:	−21	5	13	−23	−23	−23	9	12	−23	−23	15	−21	−22	−23	3	−23	13				
Classified as:	**−1**	**1**	**1**	**−1**	**−1**	**−1**	**1**	**1**	**−1**	**−1**	**1**	**−1**	**−1**	**−1**	**1**	**−1**	**1**	**17**	**0**	**0**	**100**

Note: Each gene subset is used to classify the 17 test samples. The predicted class for every sample is represented by a vote value (1 for normal, −1 for tumor, and 0 for unknown).

Table 5.9. Optimal gene subsets and selected genes for the leukemia data class predictor.

Gene subsets	Gene IDs																						
1-gene subset	5501																						
2-gene subset	3320	1068																					
3-gene subset	2020	4782	2348																				
4-gene subset	4270	2039	4050	2642																			
5-gene subset	2642	1837	4050	1488	5605																		
6-gene subset	3137	2642	3336	2368	2852	4050																	
7-gene subset	2020	1725	2531	2096	4991	2348	2120																
8-gene subset	2642	4492	307	6368	3753	4708	5655	4050															
9-gene subset	1481	4991	2224	2642	109	4050	5094	3565	6441														
10-gene subset	2619	3119	3056	2971	4339	5297	2861	2020	5247	2001													
11-gene subset	1584	4023	2020	1506	2852	4459	1060	6467	2295	2348	2483												
12-gene subset	6910	4669	2642	6939	1891	4050	2020	4916	6487	1442	4950	2128											
13-gene subset	1934	3906	3010	3392	5906	7129	4453	4724	4961	2280	2642	1	4050										
14-gene subset	1362	2642	6771	4050	2090	6681	2811	988	4574	4727	5673	3191	1427	3565									
15-gene subset	2020	1853	501	1387	4414	3565	3056	1630	6243	1143	2342	5251	4139	4720	1834								
16-gene subset	4751	6438	3414	4224	5949	4889	1056	6559	2642	2648	5210	5166	1	4050	3888	5134							
17-gene subset	7129	1834	4640	3189	6872	3118	2433	4050	1740	5326	4768	2469	6042	1	4444	2642	4252						
18-gene subset	5828	442	3299	6548	2400	2378	3525	5452	4127	2642	5770	5342	6319	1945	4050	2780	6136	4464					
19-gene subset	714	1714	5912	4711	3839	3215	2506	2642	3804	1900	5299	5609	4050	1	6655	6372	2791	1211	3068				
20-gene subset	3515	7129	3854	6762	5826	4050	1250	2416	1021	3322	5451	5508	4410	2642	2327	4037	6639	4278	4334	5745			
21-gene subset	92	6833	2642	1385	1801	3102	4251	4050	6832	5651	2449	4189	1925	5826	301	1126	3034	6940	1594	3342	5384		
22-gene subset	5064	5692	6034	4050	6435	2642	628	501	4960	4908	5882	2227	3998	2004	4723	7021	1	2829	1513	3423	3642	2642	
23-gene subset	7129	5477	714	4534	4572	643	3066	4991	2327	1229	1425	4634	3565	4634	3565	4416	3149	4452	1250	4063	3026	2642	3780

Table 5.10. Validation of the leukemia data with the multi-subset class predictor.

Gene subsets	Prediction of 34 test samples (No. 39–No. 72)																																		Results of Classification			
	39	40	41	42	43	44	45	46	47	48	49	50	51	52	53	54	55	56	57	58	59	60	61	62	63	64	65	66	67	68	69	70	71	72	Correct	Incorrect	Unknown	Success rate %
1-gene subset	+1	0	0	0	+1	+1	+1	+1	+1	+1	+1	0	0	0	0	0	0	+1	0	0	+1	0	0	0	0	+1	-1	0	+1	+1	+1	0	0	+1	18	2	14	52.9
2-gene subset	+1	+1	+1	+1	+1	+1	+1	+1	+1	+1	+1	0	-1	0	-1	0	+1	+1	0	-1	+1	+1	+1	-1	-1	-1	+1	+1	+1	+1	+1	+1	+1	+1	27	5	2	79.4
3-gene subset	+1	+1	-1	+1	+1	+1	+1	+1	+1	+1	+1	-1	+1	+1	-1	-1	+1	+1	0	-1	+1	0	0	-1	0	-1	+1	+1	+1	+1	+1	+1	+1	+1	25	5	4	73.5
4-gene subset	+1	0	+1	+1	+1	+1	+1	+1	+1	+1	+1	-1	-1	+1	-1	+1	0	+1	-1	-1	+1	0	-1	0	-1	-1	+1	+1	+1	+1	+1	+1	+1	+1	31	0	3	91.2
5-gene subset	+1	0	+1	+1	+1	+1	+1	+1	+1	+1	+1	-1	-1	-1	-1	-1	0	+1	-1	-1	+1	-1	-1	-1	-1	-1	+1	+1	+1	+1	+1	+1	+1	+1	31	0	3	91.2
6-gene subset	+1	-1	+1	+1	+1	+1	+1	+1	+1	+1	+1	-1	-1	+1	-1	-1	0	+1	-1	-1	0	-1	-1	-1	-1	-1	+1	0	+1	+1	+1	+1	+1	+1	30	1	3	88.2
7-gene subset	+1	+1	0	+1	+1	+1	+1	+1	+1	+1	+1	-1	+1	+1	-1	0	+1	+1	0	-1	+1	0	0	-1	-1	-1	+1	+1	+1	+1	+1	+1	+1	+1	27	4	3	79.4
8-gene subset	+1	+1	+1	+1	+1	+1	+1	+1	+1	+1	+1	-1	-1	+1	-1	-1	+1	+1	0	-1	+1	-1	-1	0	-1	-1	+1	0	+1	+1	0	+1	+1	+1	29	3	2	85.3
9-gene subset	+1	0	+1	+1	+1	+1	+1	+1	+1	+1	+1	-1	-1	-1	-1	0	+1	+1	+1	-1	+1	0	-1	-1	-1	-1	+1	+1	+1	+1	0	+1	+1	+1	31	0	3	91.2
10-gene subset	+1	+1	0	+1	+1	+1	+1	+1	+1	+1	+1	-1	+1	+1	-1	+1	+1	+1	0	-1	+1	0	0	-1	+1	-1	+1	+1	+1	+1	+1	+1	+1	+1	24	8	2	70.6
11-gene subset	+1	+1	-1	+1	+1	+1	+1	+1	+1	+1	+1	-1	+1	+1	-1	+1	0	+1	+1	-1	+1	+1	-1	-1	+1	-1	+1	+1	+1	+1	+1	+1	+1	+1	26	6	2	76.4
12-gene subset	+1	+1	+1	+1	+1	+1	+1	+1	+1	+1	+1	-1	-1	-1	-1	0	+1	+1	-1	-1	+1	-1	-1	-1	-1	-1	+1	+1	+1	+1	+1	+1	+1	+1	32	0	2	94.1
13-gene subset	+1	0	+1	+1	+1	+1	+1	+1	+1	+1	+1	-1	-1	-1	-1	0	+1	+1	-1	-1	+1	0	-1	-1	-1	-1	+1	+1	+1	+1	+1	+1	+1	+1	31	0	3	91.2
14-gene subset	+1	0	+1	+1	+1	+1	+1	+1	+1	+1	+1	-1	-1	-1	-1	+1	0	+1	-1	-1	+1	0	-1	-1	-1	-1	+1	+1	+1	+1	+1	+1	+1	+1	31	0	3	91.2
15-gene subset	+1	+1	0	+1	+1	+1	+1	+1	+1	+1	+1	0	-1	+1	-1	+1	+1	+1	0	-1	+1	-1	-1	-1	-1	+1	+1	+1	+1	+1	+1	+1	+1	+1	27	4	3	79.4
16-gene subset	+1	0	+1	+1	+1	+1	+1	+1	+1	+1	+1	-1	-1	-1	-1	0	+1	+1	-1	-1	+1	-1	-1	-1	-1	-1	+1	+1	+1	+1	+1	+1	+1	+1	31	0	3	91.2
17-gene subset	+1	0	+1	+1	+1	+1	+1	+1	+1	+1	+1	-1	-1	-1	-1	0	+1	+1	-1	-1	+1	-1	-1	-1	-1	-1	+1	+1	+1	+1	+1	+1	+1	+1	32	0	2	94.1
18-gene subset	+1	0	+1	+1	+1	+1	+1	+1	+1	+1	+1	-1	-1	-1	-1	0	+1	+1	-1	-1	+1	-1	-1	-1	-1	-1	+1	+1	+1	+1	+1	+1	+1	+1	32	0	2	94.1
19-gene subset	+1	0	+1	+1	+1	+1	+1	+1	+1	+1	+1	-1	-1	-1	-1	0	+1	+1	-1	-1	+1	0	-1	-1	-1	-1	+1	+1	+1	+1	+1	+1	+1	+1	31	0	3	91.2
20-gene subset	+1	0	+1	+1	+1	+1	+1	+1	+1	+1	+1	-1	-1	-1	-1	0	+1	+1	-1	-1	+1	0	-1	-1	-1	-1	+1	+1	+1	+1	+1	+1	+1	+1	31	0	3	91.2
21-gene subset	+1	0	+1	+1	+1	+1	+1	+1	+1	+1	+1	-1	-1	-1	-1	0	+1	+1	-1	-1	+1	0	-1	-1	-1	-1	+1	0	+1	+1	+1	+1	+1	+1	31	0	3	91.2
22-gene subset	+1	0	+1	+1	+1	+1	+1	+1	+1	+1	+1	-1	-1	-1	-1	0	+1	+1	-1	-1	+1	0	-1	-1	-1	-1	+1	0	+1	+1	+1	+1	+1	+1	31	0	3	91.2
23-gene subset	+1	0	+1	+1	+1	+1	+1	+1	+1	+1	+1	-1	-1	-1	-1	-1	0	+1	-1	-1	+1	-1	-1	-1	-1	-1	+1	0	+1	+1	+1	+1	+1	+1	30	0	4	88.2
Sum of Votes:	+23	+7	+23	+15	+23	+23	+23	+23	+23	+23	+22	-22	-10	-22	-17	+7	+23	+23	-15	-23	+22	-8	-15	-4	-19	-21	-13	-8	+20	+22	+23	+22	+22	+23	34	0	0	100
Classified as:	+1	+1	+1	+1	+1	+1	+1	+1	+1	+1	+1	-1	-1	-1	-1	+1	+1	+1	-1	-1	+1	-1	-1	-1	-1	-1	-1	-1	+1	+1	+1	+1	+1	+1				

Note: Each gene subset is used to classify the 34 test samples. The predicted class for every sample is represented by a vote value (+1 for ALL, −1 for AML, and 0 for unknown).

is assigned to the votes, and then the sum of the votes $\{11 \times (+1) + 6 \times (-1) + 6 \times 0 = +5\}$ is calculated. Since the sum is +5, which means there are more votes favoring the normal class, sample N28 is classified as "normal" and is given a class code of +1. Otherwise, the sample should be classified as "tumor" with a code of −1 if the sum of the votes is a negative value. The absolute value of the sum (ranging from 1 to 23) indicates the predicting strength. When the sum equals zero, the sample under test is unclassifiable.

For the sum of votes in the colon cancer data, one adds up the vote values across all 23 gene subsets for every sample. If the sum of the 23 prediction votes has a positive value (which indicates that there are more gene subsets favoring the normal class than the tumor class), the corresponding sample is classified as a normal sample with a code of 1. If the sum of the 23 prediction votes has a negative value (which indicates that there are more gene subsets favoring the tumor class than the normal class), the corresponding sample is classified as a tumor sample with a code of −1. Where the sum of the 23 prediction votes equals zero, the corresponding sample is unclassified with a code of 0.

For the sum of votes in the leukemia data, one adds up the vote values across all 23 gene subsets for every sample. If the sum of the 23 prediction votes is positive (which indicates that there are more gene subsets favoring the ALL class than the AML class), the corresponding sample is classified as an ALL sample with a code of 1. If the sum of the 23 prediction votes is negative (which indicates that there are more gene subsets favoring the AML class than the ALL class), the corresponding sample is classified as an AML sample with a code of −1. Where the sum of the 23 prediction votes equals zero, the corresponding sample is unclassified with a code of 0.

5.5. Discussion

Optimization algorithms are playing a significant role in the field of gene selection and sample classification for microarray data. Many advanced local and global optimization techniques, such as clustering and genetic algorithms, have been successfully applied to gene subset selection for classifying cancer tissue samples. Any optimization algorithm applied to a particular selection problem should first address the issue of choosing a reasonable starting solution, which is always a big obstacle for an inexperienced operator. To find the true global optimized solution for a gene selection problem, one needs to solve an array of interlinked multi-dimensional simultaneous equations. For a gene subset with more than just a few gene elements, until recently this has been a very difficult task, requiring the use of a supercomputer and highly skilled programming. With the help of SDL global optimization, however, there is no need to solve equations. The global optimized solutions can be found at an affordable computing cost through orthogonal sampling.

It is worth observing that the established multi-subset class predictor could be reduced in size by removing the first five or more unstable short gene subsets; the remaining subsets would still perform well, as shown on the supporting website (Li, 2006). In general, the predicting strength may be improved. However, having those genes selected in the short subsets included may be significant to biologists, as they could well be informative.

Another interesting observation is that there are not many genes which each plays a more important role than any other gene. The most frequently appearing genes involved in the colon cancer predictor were 249 and 164, which appeared 10 times and eight times, respectively. Most of the genes in the predictor were selected only once. For the leukemia predictor, the situation is quite similar. Genes 2642 and 4050 were the most frequently used genes, being included 16 times each. Both gene IDs assigned by this study and real gene accession numbers from the original datasets are listed in Tables 5.11 and 5.12, respectively. The gene appearance frequency for the colon class predictor is also given in Table 5.11.

Some previous research works proposed to find out many near-optimal gene subsets through a well-tuned GA procedure and pick up the top 50–200 most frequently appearing genes to construct a long gene subset as a predictor (Li *et al.*, 2001b). Although the performance of such a predictor was reasonably good, the large amount of computation might not be affordable or cost-effective and might not be necessary. One more experiment was quickly carried out by forming a subset with the seven most frequently appearing genes identified from the colon cancer predictor; they are genes 249, 164, 2000, 245, 567, 66, and 581. Using such

Table 5.11.   The 221 genes selected by the colon class predictor and their appearance frequency.

IDs	Genes	Freq.	IDs	Genes	Freq.	IDs	Genes	Freq.	IDs	Genes	Freq.	IDs	Genes	Freq.	IDs	Genes	Freq.	IDs	Genes	Freq.
66	T71025	3	479	U20428	1	757	T79595	1	1025	T74896	1	1231	H49870	1	1497	R00254	1	1706	H71627	1
118	T72889	1	493	R87126	1	760	M36341	1	1033	M69066	1	1247	X74295	1	1519	X69115	1	1714	X79888	1
138	M26697	1	504	X55362	1	765	M76378	3	1041	R54467	1	1263	T40454	1	1538	R67987	1	1735	H62885	1
164	X57351	8	508	H01677	1	767	U07695	1	1042	R36977	2	1264	H47107	1	1539	H78346	1	1761	T94350	1
206	L16510	1	510	D90188	1	773	R40184	1	1043	M86737	2	1272	M92843	1	1550	X53799	1	1786	M22050	1
228	J03040	2	513	M22382	2	778	X69550	1	1058	M80815	2	1282	L25081	1	1551	U20141	1	1792	M85169	1
239	H48027	1	539	H65182	1	782	X02750	1	1066	R83349	1	1286	D16294	2	1557	M21186	1	1800	U18062	2
240	T40507	1	541	X78706	1	802	X70326	2	1067	T70062	1	1294	U21049	1	1558	R49416	1	1802	T40674	1
245	M76378	5	542	M13686	1	804	R76254	1	1070	R70535	1	1312	M86934	1	1562	R49459	1	1805	H82741	1
247	T79813	1	547	M37984	1	813	M83738	1	1071	H40108	1	1327	T84082	1	1582	X63629	1	1809	H24310	1
249	M63391	10	550	M94630	1	831	T53868	1	1085	D42053	1	1351	X68688	1	1589	R71651	1	1812	X93510	1
251	U37012	4	567	X65488	1	853	H15813	2	1086	R39531	1	1360	H09719	2	1596	X51435	1	1818	H40699	1
254	H80975	1	572	T58756	1	855	L37033	1	1087	T55117	1	1362	M95549	2	1600	M90516	3	1819	T71049	2
264	M26252	1	581	T51571	3	857	R43976	1	1095	H04282	1	1366	Z11502	1	1608	L12723	1	1840	D38537	1
276	T53694	1	583	H59599	1	871	J03069	1	1102	T51558	1	1370	H22579	1	1610	U15782	1	1842	M87434	1

(Continued)

Table 5.11. *(Continued)*

IDs	Genes	Freq.	IDs	Genes	Freq.	IDs	Genes	Freq.	IDs	Genes	Freq.	IDs	Genes	Freq.	IDs	Genes	Freq.	IDs	Genes	Freq.
288	T48102	2	586	H17897	1	883	H04235	1	1103	T56460	1	1372	X75208	1	1611	H92195	1	1846	T52806	1
295	X15183	1	594	T89422	1	885	H27202	1	1110	L08069	1	1380	M92287	1	1623	T94993	1	1849	V00520	1
309	T70331	1	599	T53412	1	888	H89688	1	1128	D11466	1	1393	H77536	1	1626	T95612	1	1850	M22760	1
316	U26401	1	605	J03075	1	897	H43887	2	1138	T56475	2	1402	M38690	1	1644	R80427	1	1863	X89985	1
323	U28963	1	619	H89087	1	898	J02906	1	1165	H63354	1	1412	X61118	1	1656	X16504	1	1873	L07648	1
327	U37408	1	623	H11324	1	902	R00544	1	1166	R44740	1	1415	X12548	1	1657	R99935	1	1880	R09502	1
340	T87527	1	629	T60318	1	904	M64929	1	1171	U23852	1	1423	J02854	2	1662	H08751	1	1890	M19156	1
341	D13315	1	633	H87344	1	911	K02566	1	1173	M55543	1	1424	D31887	1	1663	X16663	1	1905	R23907	1
344	H15542	1	642	X16316	1	924	U09848	1	1174	R60906	1	1430	R09468	1	1665	X59842	1	1908	L25851	1
355	M30474	1	662	X68277	1	948	H45299	1	1196	D13315	1	1436	D10523	1	1670	H23975	2	1915	H87193	1
362	M36205	1	665	R38513	1	977	L40904	1	1201	X75304	1	1437	T67433	1	1672	D42047	1	1943	D29808	1
394	M99564	1	679	X70944	1	980	U06698	1	1208	H72965	1	1445	M13560	1	1674	T67077	1	1945	M69181	1
401	L19956	1	693	L11370	1	990	D38549	1	1209	R96357	1	1461	L34840	1	1676	U03851	1	1970	R06749	1
417	R61332	1	698	T51261	2	994	T51858	1	1213	M93651	1	1469	R38736	1	1678	R96070	1	2000	T49647	6
422	R77824	1	712	T90036	1	996	X79683	1	1216	R61324	1	1472	L41559	1	1680	M31516	1			
457	H22939	1	723	H65019	1	1009	X86018	1	1217	T62067	1	1487	L25941	1	1686	L47162	1			
471	J04173	1	741	T77446	1	1020	X56253	2	1227	T96873	2	1490	L00354	1	1699	R39540	1			

Table 5.12.　The 219 genes selected by the leukemia class predictor.

IDs	Gene accession number	IDs	Gene accession number	IDs	Gene accession number	IDs	Gene accession number	IDs	Gene accession number
5501	Z15115	1506	L36051	2342	M90696	6136	U28749_s	1925	M31165
3320	U50136_rna1	4459	X67683	5251	D28791	4464	X68149	301	D25303
1068	J03040	1060	J02883	4139	X13956	714	D87443	1126	J04809_rna1
2020	M55150	6467	U29463_s	4720	X85134_rna1	1714	M14123_xpt2	3034	U31449
4782	X90908	2295	M85169	1834	M23197	5912	HG880-HT880	6940	Z30644
2348	M91432	2483	S73813	4751	X87342	4711	X84195	1594	L41147
4270	X54936	6910	U84388	6438	S77154_s	3839	U82320	3342	U51166
2039	M57471	4669	X81889	3414	U56814	3215	U43522	5384	U13022
4050	X03934	6939	Z30643	4224	X52001	2506	S77576	5064	Z15108
2642	U05259_rna1	1891	M28713	5949	M29610	3804	U80017_rna2	5692	D89377_s
1837	M23379	4916	X99657	4889	X98263	1900	M29273	6034	U50360_s
1488	L34357	6487	X75346_s	1056	J02843	5299	L07919	6435	U05012_s
5605	D29675	1442	L27479	6559	U41315_rna1_s	5609	X14085_s	628	D83784
3137	U38846	4950	Y07596	2648	U05875	6655	Z11518_s	4960	Y07846
3336	U50939	2128	M63379	5210	Z79581	6372	M81182_s	4908	X99268

(Continued)

Table 5.12. (*Continued*)

IDs	Gene accession number	IDs	Gene accession number	IDs	Gene accession number	IDs	Gene accession number	IDs	Gene accession number
2368	M93284	1934	M31642	5166	Z48804	2791	U14550	5882	HG417-HT417_s
2852	U18004	3906	U89278	3888	U86782	1211	L05512	2227	M76558
1725	M14636	3010	U30245	5134	Z35491	3068	U33818	3998	U96629_rma2
2531	S81221	3392	U53476	1834	M23197	3515	U62437	2004	M37763
2096	M61156	5906	X07618_s	4640	X80062	3854	U83303_cds2	4723	X85372
4991	Y09615	7129	Z78285_f	3189	U41813	6762	M21388	7021	M33318_r
2120	M62994	4453	X67155	6872	M92642	5826	HG3125-HT3301_s	2829	U16296
4492	X69908_rma1	4724	X85373	3118	U37283	1250	L08424	1513	L36645
307	D26067	4961	Y07847	2433	S34389	2416	M97639	3423	U57099
6368	M80397_s	2280	M83651	1740	M15841	1021	HG511-HT511	3642	U70732_rma1
3753	U79249	1	AFFX-BioB-5	5326	M13577	3322	U50315	5477	X71661
4708	X84002	1362	L19067	4768	X89750	5451	X14766	4534	X74104
5655	U58046_s	6771	X87344_cds10_r	2469	S70348	5508	HG2157-HT2227	4572	X76105
1481	L33881	2090	M60749	6042	L10333_s	4410	X64643	643	D85376
2224	M76424	6681	X74874_rma1_s	4444	X66534	2327	M88282	3066	U33447

(*Continued*)

Table 5.12. *(Continued)*

IDs	Gene accession number	IDs	Gene accession number	IDs	Gene accession number	IDs	Gene accession number	IDs	Gene accession number
109	AC002115_cds4	2811	U15177	4252	X53742	4037	X02751	1229	L07077
5094	Z24727	988	HG4245-HT4515	5828	HG3187-HT3366_s	6639	U83598	1425	L25270
3565	U66048	4574	X76180	442	D45370	4278	X55666	4634	X79865
6441	S78873_s	4727	X85750	3299	U49187	4334	X59711	6416	S57153_s
2619	U03644	5673	D85425_s	6548	Z69030_s	5745	HG2261-HT2351_s	4452	X67098
3119	U37352	3191	U41816	2400	M95925	92	AB003698	3149	U39412
3056	U32944	1427	L25444	2378	M94167	6833	J00220_cds5	4063	X04434
2971	U27185	3565	U66048	3525	U63289	1385	L20348	3026	U31120_rna1
4339	X59812	1853	M25077	5452	X15422	1801	M21154	3780	U79287
5297	L07615	501	D50931	4127	X12901	3102	U36501		
2861	U18288	1387	L20773	5770	X52009_s	4251	X53587		
5247	D17532	4414	X64838	5342	M37712	6832	J00210_rna1		
2001	M37435	1630	L47738	6319	M60450_s	5651	D50477_s		
1584	L40410	6243	M24486_s	1945	M32315	2449	S76992		
4023	X01059	1143	J05213	2780	U13737	4189	X16667		

a seven-gene subset to classify the samples in the test dataset, the results were very encouraging. All of the 17 samples were correctly classified (data not shown here), which indicates that selecting the most frequently appearing genes to form a subset for classification may gain significant advantages, although these genes may show more biological significance. The more important point is that the length of the gene subset can be shortened greatly from around 100 genes to around 10 by SDL.

For comparison with the outputs from the GA algorithms (Li *et al.*, 2001a), the validation experiments were carried out further. There are two genes appearing most frequently in the colon class predictor: gene 249 (10 times) and gene 164 (8 times). A subset with only these two genes is able to classify 16 out of 17 samples in the colon cancer test dataset, while one sample remains as unclassifiable. Although gene 164 (X57351) was included in the 50 genes identified by a GA class predictor, the most frequently selected gene 249 (M63391) in this research was not captured before. The most frequently selected gene by GA was the human monocyte-derived neutrophil-activating protein (MONAP). Previous studies have demonstrated that the expression level of the MONAP gene, whose gene ID is 1671 in this research, directly correlates with the progression of several human cancers (Shi *et al.*, 1999). Unfortunately, gene 1671 was missed completely by the SDL method, which might reflect the fact that there are fundamental differences between GA and SDL in terms of sampling the search spaces to solve the problems.

For the leukemia data, 219 (shown in Table 5.4) out of 7129 genes in the dataset were selected by SDL for constructing the class predictor.

Table 5.13 lists the genes appearing more than once in the leukemia class predictor based on frequency rank. It is worthwhile to note that gene 2642 (U05259_ma1) and gene 4050 (X03934) both appear 16 times. A subset with only these two genes is able to classify 31 out of 34 samples in the leukemia test dataset (three samples remain unclassified). When a subset of the top four genes (2642, 4050, 2020, and 1) is used, 32 out of 34 samples can be predicted correctly, with two remaining unclassified. There are four genes (2642, 2020, 2348, and 3056) in Table 5.13 that were identified by previous researchers as among the 50 genes most highly correlated with the ALL–AML class distinction (Golub *et al.*, 1999); with the method used in this study, 29 out of

Table 5.13.   **Genes that appear more than once in the leukemia class predictor.**

Rank	Gene IDs	Frequency	Gene accession number	Gene description
1	2642	16	U05259_rna1	MB-1 gene
2	4050	16	X03934	GB DEF = T-cell antigen receptor gene T3-delta
3	2020	6	M55150	FAH Fumarylacetoacetate
4	1	6	AFFX-BioB-5	AFFX-BioB-5_at (endogenous control)
5	3565	4	U66048	Clone 161455 breast expressed mRNA from chromosome X
6	7129	4	Z78285_f	GB DEF = mRNA (clone 1A7)
7	2348	3	M91432	ACADM Acyl-Coenzyme A dehydrogenase, C-4 to C-12 straight chain
8	4991	3	Y09615	GB DEF = Mitochondrial transcription termination factor
9	2852	2	U18004	HSU18004 *Homo sapiens* cDNA
10	3056	2	U32944	Cytoplasmic dynein light chain 1 (hdlc1) mRNA
11	5826	2	HG3125-HT3301_s	Estrogen receptor (Gb:S67777)
12	501	2	D50931	KIAA0141 gene
13	714	2	D87443	KIAA0254 gene
14	2327	2	M88282	T-cell surface protein tactile precursor
15	1250	2	L08424	Achaete scute homologous protein (ASH1) mRNA

34 test samples could be classified correctly. In contrast, the SDL global optimization method classified all of them correctly. Moreover, the SDL method, by selecting sets of genes based on their joint ability to discriminate, can identify genes that are important jointly, but do not discriminate individually. This indicates that the SDL method has potential in identifying genes that not only discriminate between ALL and AML, but also distinguish existing subtypes without applying any prior knowledge.

5.6. Conclusion

DNA microarrays make it practical, for the first time, to survey the expression of thousands of genes under thousands of conditions. This technology makes it possible to study the expression of all of the genes at once. Large-scale expression profiling has emerged as a leading technology in the systematic analysis of cellular physiology. However, method development for analyzing gene expression data is still in its infancy. SDL optimization uses a mathematical method based on orthogonal sets of numbers. By slicing the multi-dimensional parameter space with a horizontal plane of the objective function, with each parameter independent of the others, a peak is always surrounded by a slope. By finding all regions in which the objective function has values above that of the plane, one can narrow the search region. After finding the boundary of all the isolated regions where this occurs, the plane is raised again, and the process repeated.

Orthogonal arrays (OAs) are immensely important in all areas of human investigation. In statistics, they are primarily used in designing experiments. An OA is an array of numbers constructed by utilizing orthogonal Latin squares. One can form an array of several dimensions that are orthogonal to each other, and therefore allow the calculation of a resultant using many interdependent variables. Combining OA sampling with function domain contraction techniques results in an optimization with two desirable properties. Firstly, the number of function evaluations can be greatly reduced. Secondly, there is a guarantee of finding the global optimum solution. In this study, a carefully selected OA was successfully used for conducting an orthogonal search space sampling.

By using OA and other mathematical techniques, it is practical to develop a global optimization program for cancer classification and validation on a desktop computer. The primary advantages of this technique are that the global optimum is always found, excellent solutions can be found with little prior knowledge, and the new objective functions can be created according to whatever combination of parameters is required. The mathematical procedures used in this form of global optimization are possible to apply to a variety of other previously unsolved problems relating to the resultant of dependent variables, including experimental design and manufacturing variations. Many other approaches have been adopted, but until now (with the exception of scanning) they all depend either on a starting design, some form of local optimization, or some random variation. Each method will usually give rise to different solutions. For gene subsets using a large number of genes, these are still the only methods possible. In contrast, the SDL optimization described here is a methodical global method.

The proposed pyramidal hierarchy of the predictor for classification can effectively improve the signal-to-noise ratio in mining the high-dimensional microarray datasets. While research in cancer classification with microarray expression data is the first to benefit from this method, the mathematical procedures and SDL global optimization used in this study are also applicable to a variety of other unsolved problems related to linked multi-variable problems. The application of this technique will undoubtedly have implications well beyond cancer classification application.

It is still too early to predict what the ultimate impact of microarray will be on our understanding of cancer, although the possibility of an accurate diagnosis of cancers based on microarray expressions has emerged. This innovative research truly brings to light one of the hardest problems yet — the ability to accurately classify medical neoplasm. The SDL method provides a precise diagnostic tool that can find the true global optima with questions relating to gene malignancy. Furthermore, genetic screening for diseases is playing an increasingly important role in preventative medicine. If we can detect the presence of disease or predict malignancy through microarray expression data with a desktop computer before clinical diagnosis, a more efficient and clear-cut treatment plan can be formulated, eliminating the possibility of clinician bias. More importantly, an unbiased and digital data-based approach can be easily applied to distinctions relating to future clinical outcome, such as drug response or survival. In cancer research, fundamental mechanisms that cut across distinct types of cancers could also be discovered through mining microarray data by the SDL global strategies.

CHAPTER 6

General Discussion and Future Directions

As emphasized in Chapter 1, DNA microarray technology allows the parallel and simultaneous detection of more than 30000 genes in cells. Although large genome-scale cDNA or RNAi screens are powerful and efficient, they examine only one gene at a time, and will not uncover biological activities that often rely on multiple collaborating genes. Thus, DNA microarray is one of the best assays available for studying complex biological processes at a transcriptional level. At present, DNA microarray technology in cancer research is at a stage that, at best, provides information about disease-associated molecular signatures derived from analysis of the expression of basically all genes. There is still a long way to go before a diagnostic decision can be made based solely on DNA microarray data. This is largely due to the lack of a more powerful method for analysis of DNA microarray data. The development of new analytical methods of DNA microarray data is a critical step to take. In this regard, our SDL global optimization method described in this book has significantly changed our way of analyzing DNA microarray data, and has brought us much closer to making decisions based on microarray data.

As we know, a gene is a fundamental constituent of any living organism. The machinery of each human body is built and run with 50000 to 100000 different kinds of genes or protein molecules. With the completion of the Human Genome Project, one has access to large databases of biological information. Proper analysis of such huge data holds immense promise in bioinformatics. The applicability of data mining in this domain cannot be denied, given the life-saving prospects of effective drug design. This is also of practical interest to the pharmaceutical industry (Mitra and Acharya, 2005).

The success of DNA technologies and the digital revolution brought about by the growth of the Internet have ensured that huge volumes of high-dimensional microarray expression data are highly available. Data mining is an evolving and growing area of research and development. The problem is to mine useful information or patterns from the huge datasets. Microarrays provide a powerful basis to monitor the expression of tens of thousands of genes in order to identify mechanisms that govern the activation of genes in an organism. Microarray experiments are done to produce gene expression patterns, which provide dynamic information about cell function. The huge volume of such data, and their high dimensions, make gene expression data suitable candidates for the application of data mining functions.

We have provided an introduction to knowledge discovery from microarray experimental datasets. The major functions of data mining have been discussed from the perspectives of machine learning, pattern recognition, and artificial intelligence.

Soft computing methodologies, involving fuzzy sets, neural networks, genetic algorithms (GAs), rough sets, wavelets, and their hybridizations, have recently been used to solve data mining problems. They strive to provide approximate solutions at low cost, thereby speeding up the process. For future research and development in microarray data analysis and mining, all of these methods are useful tools. The role of soft computing in microarray gene expression study is very promising with the learning ability of neural networks to predict, the searching potential of GAs and SDL, and the uncertainty-handling capacity of fuzzy sets. On the other hand, by using an orthogonal array (OA) and other mathematical techniques, it is practical to develop a global optimization program for cancer classification and validation on a desktop computer. The primary advantages of this technique are that the global optimum is always found, excellent solutions can be found with little prior knowledge, and the new objective functions can be created according to whatever combination of parameters is required.

The mathematical procedures used in this form of global optimization are possible to apply to a variety of other previously unsolved problems relating to the resultant of dependent variables, including experimental design and manufacturing variations. There are many other approaches that people have adopted, but until now (with the exception of scanning) they all depend either on a starting design, some form of local optimization, or some random variation. Each method will usually give rise to different solutions. For gene subsets using a large number of genes, these are still the only methods possible. In contrast, the SDL optimization described here is a methodical global method.

The proposed pyramidal hierarchy of the predictor for classification can effectively improve the signal-to-noise ratio in mining the high-dimensional microarray datasets. While research in cancer classification with microarray expression data is the first to benefit from this method, the mathematical procedure used in this study — SDL global optimization — is also applicable to a variety of other unsolved problems related to linked multi-variable problems. The application of this technique will undoubtedly have implications well beyond cancer classification application.

It is still too early to predict what the ultimate impact of microarray will be on our understanding of cancer, although the possibility of an accurate diagnosis of cancers based on microarray expressions has emerged. This innovative research truly brings to light one of the hardest problems yet: the ability to accurately classify medical neoplasm. The SDL method provides a precise diagnostic tool that can find the true global optima with questions relating to gene malignancy. Furthermore, genetic screening for diseases is playing an increasingly important role in preventative medicine — if we can detect the presence of disease or predict the malignancy through microarray expression data with a desktop computer before clinical diagnosis, a more efficient and clear-cut treatment plan can be formulated, eliminating the possibility of clinician bias. More importantly, an unbiased and digital data-based approach can be easily applied to distinctions relating to future clinical outcome, such as drug response or survival. In cancer research, fundamental mechanisms that cut across distinct types of cancers could also be discovered through mining microarray data using SDL global strategies. In the future, the use of our SDL global optimization method in DNA microarray data analysis has great potential to help simultaneously gather disease information related to the expression of all genes from a patient — i.e. personalized medicine.

References

[Anonymous.] (2002). Special Issue on Bioinformatics. *IEEE Comput* **35**.

[Anonymous.] (2006). Making the most of microarrays. *Nat Biotechnol* **24**: 1039.

Abruzzo, L.V., Wang, J., Kapoor, M., Medeiros, L.J., Keating, M.J., Highsmith, W.E., Barron, L.L., Cromwell, C.C., and Coombes, K.R. (2005). Biological validation of differentially expressed genes in chronic lymphocytic leukemia identified by applying multiple statistical methods to oligonucleotide microarrays. *J Mol Diagn* **7**: 337–345.

Abul, O., Alhajj, R., Polat, F., and Barker, K. (2005). Finding differentially expressed genes for pattern generation. *Bioinformatics* **21**: 445–450.

Adessi, C., Matton, G., Ayala, G., Turcatti, G., Mermod, J.J., Mayer, P., and Kawashima, E. (2000). Solid phase DNA amplification: characterisation of primer attachment and amplification mechanisms. *Nucleic Acids Res* **28**: E87.

Advani, A.S., and Pendergast, A.M. (2002). Bcr-Abl variants: biological and clinical aspects. *Leuk Res* **26**: 713–720.

Aittokallio, T., Kurki, M., Nevalainen, O., Nikula, T., West, A., and Lahesmaa, R. (2003). Computational strategies for analyzing data in gene expression microarray experiments. *J Bioinform Comput Biol* **1**: 541–586.

Alizadeh, A.A., Eisen, M.B., Davis, R.E., Ma, C., Lossos, I.S., Rosenwald, A., Boldrick, J.C., Sabet, H., Tran, T., Yu, X., *et al.* (2000). Distinct types of diffuse large B-cell lymphoma identified by gene expression profiling. *Nature* **403**: 503–511.

Allison, D.B., Cui, X., Page, G.P., and Sabripour, M. (2006). Microarray data analysis: from disarray to consolidation and consensus. *Nat Rev Genet* **7**: 55–65.

Alon, U., Barkai, N., Notterman, D.A., Gish, K., Ybarra, S., Mack, D., and
Levine, A.J. (1999). Broad patterns of gene expression revealed by clustering analysis of tumor and normal colon tissues probed by oligonucleotide
arrays. *Proc Natl Acad Sci USA* **96**: 6745–6750.

Amarante-Mendes, G.P., Kim, C.N., Liu, L., Huang, Y., Perkins, C.L., Green,
D.R., and Bhalla, K. (1998). Bcr-Abl exerts its antiapoptotic effect against
diverse apoptotic stimuli through blockage of mitochondrial release of
cytochrome C and activation of caspase-3. *Blood* **91**: 1700–1705.

Anderle, P., Duval, M., Draghici, S., Kuklin, A., Littlejohn, T.G., Medrano, J.F.,
Vilanova, D., and Roberts, M.A. (2003). Gene expression databases and
data mining. *Biotechniques* (Suppl): S36–S44.

Anderson, S.M. and Mladenovic, J. (1996). The BCR-ABL oncogene requires
both kinase activity and Src-homology 2 domain to induce cytokine secre-
tion. *Blood* **87**: 238–244.

Barrett, M.T. (2005). Stacking the chips for biological discovery. *Nat Genet*
37(Suppl): S1.

Bassett, D.E., Jr., Eisen, M.B., and Boguski, M.S. (1999). Gene expression
informatics — it's all in your mine. *Nat Genet* **21**(Suppl): 51–55.

Bergmann, S., Ihmels, J., and Barkai, N. (2003). Iterative signature algorithm for
the analysis of large-scale gene expression data. *Phys Rev E Stat Nonlin Soft
Matter Phys* **67**: 031902.

Bianchi, F., Nuciforo, P., Vecchi, M., Bernard, L., Tizzoni, L., Marchetti, A.,
Buttitta, F., Felicioni, L., Nicassio, F., and Di Fiore, P.P. (2007). Survival
prediction of stage I lung adenocarcinomas by expression of 10 genes.
J Clin Invest **117**: 3436–3444.

Bier, F.F., Ehrentreich-Forster, E., Bauer, C.G., and Scheller, F.W. (1996a). High
sensitive competitive immunodetection of 2,4-dichlorophenoxyacetic acid
using enzymatic amplification with electrochemical detection. *Anal
Bioanal Chem* **354**: 861–865.

Bier, F.F., Ehrentreich-Forster, E., and Scheller, F.W. (1996b). Amplifying bien-
zyme cycle-linked immunoassays for determination of 2,4-dichlorophe-
noxyacetic acid. *Ann NY Acad Sci* **799**: 519–524.

Bowtell, D.D. (1999). Options available — from start to finish — for obtaining
expression data by microarray. *Nat Genet* **21**: 25–32.

Branford, S., Rudzki, Z., Walsh, S., Grigg, A., Arthur, C., Taylor, K., Herrmann, R.,
Lynch, K.P., and Hughes, T.P. (2002). High frequency of point mutations

clustered within the adenosine triphosphate-binding region of BCR/ABL in patients with chronic myeloid leukemia or Ph-positive acute lymphoblastic leukemia who develop imatinib (STI571) resistance. *Blood* **99**: 3472–3475.

Brown, M.P., Grundy, W.N., Lin, D., Cristianini, N., Sugnet, C.W., Furey, T.S., Ares, M., Jr., and Haussler, D. (2000). Knowledge-based analysis of microarray gene expression data by using support vector machines. *Proc Natl Acad Sci USA* **97**: 262–267.

Brown, P.O. and Botstein, D. (1999). Exploring the new world of the genome with DNA microarrays. *Nat Genet* **21**: 33–37.

Buckle, M., Williams, R.M., Negroni, M., and Buc, H. (1996). Real time measurements of elongation by a reverse transcriptase using surface plasmon resonance. *Proc Nat Acad Sci USA* **93**: 889–894.

Canales, R.D., Luo, Y., Willey, J.C., Austermiller, B., Barbacioru, C.C., Boysen, C., Hunkapiller, K., Jensen, R.V., Knight, C.R., Lee, K.Y., *et al.* (2006). Evaluation of DNA microarray results with quantitative gene expression platforms. *Nat Biotechnol* **24**: 1115–1122.

Casciano, D.A. and Woodcock, J. (2006). Empowering microarrays in the regulatory setting. *Nat Biotechnol* **24**: 1103.

Cheung, V.G., Morley, M., Aguilar, F., Massimi, A., Kucherlapati, R., and Childs, G. (1999). Making and reading microarrays. *Nat Genet* **21**: 15–19.

Cho, J.H., Lee, D., Park, J.H., and Lee, I.B. (2004). Gene selection and classification from microarray data using kernel machine. *FEBS Lett* **571**: 93–98.

Cho, S.B. and Ryu, J. (2002). Classifying gene expression data of cancer using classifier ensemble with mutually exclusive features. *Proc IEEE* **90**: 1744–1753.

Churchill, G.A. (2002). Fundamentals of experimental design for cDNA microarrays. *Nat Genet* **32**(Suppl): 490–495.

Cochran, W.G. and Cox, G.M. (1957). *Experimental Design*, 2nd ed. Wiley & Sons, New York.

Cui, X., Hwang, J.T., Qiu, J., Blades, N.J., and Churchill, G.A. (2005). Improved statistical tests for differential gene expression by shrinking variance components estimates. *Biostatistics* **6**: 59–75.

Danhauser-Riedl, S., Warmuth, M., Druker, B.J., Emmerich, B., and Hallek, M. (1996). Activation of Src kinases p53/56lyn and p59hck by p210bcr/abl in myeloid cells. *Cancer Res* **56**: 3589–3596.

Debouck, C., and Goodfellow, P.N. (1999). DNA microarrays in drug discovery and development. *Nat Genet* **21**: 48–50.

Deininger, M.W., Goldman, J.M., and Melo, J.V. (2000). The molecular biology of chronic myeloid leukemia. *Blood* **96**: 3343–3356.

Deutsch, J.M. (2003). Evolutionary algorithms for finding optimal gene sets in microarray prediction. *Bioinformatics* **19**: 45–52.

Dey, A., and Mukerjee, R. (1999). *Fractional Factorial Plans*. John Wiley, New York.

Dix, D.J., Gallagher, K., Benson, W.H., Groskinsky, B.L., McClintock, J.T., Dearfield, K.L., and Farland, W.H. (2006). A framework for the use of genomics data at the EPA. *Nat Biotechnol* **24**: 1108–1111.

Draghici, S. (2002). Statistical intelligence: effective analysis of high-density microarray data. *Drug Discov Today* **7**: S55–S63.

Drmanac, R., Drmanac, S., Chui, G., Diaz, R., Hou, A., Jin, H., Jin, P., Kwon, S., Lacy, S., Moeur, B., *et al.* (2002). Sequencing by hybridization (SBH): advantages, achievements, and opportunities. *Adv Biochem Eng Biotechnol* **77**: 75–101.

Druker, B.J., Sawyers, C.L., Kantarjian, H., Resta, D.J., Reese, S.F., Ford, J.M., Capdeville, R., and Talpaz, M. (2001a). Activity of a specific inhibitor of the BCR-ABL tyrosine kinase in the blast crisis of chronic myeloid leukemia and acute lymphoblastic leukemia with the Philadelphia chromosome. *N Engl J Med* **344**: 1038–1042.

Druker, B.J., Talpaz, M., Resta, D.J., Peng, B., Buchdunger, E., Ford, J.M., Lydon, N.B., Kantarjian, H., Capdeville, R., Ohno-Jones, S., *et al.* (2001b). Efficacy and safety of a specific inhibitor of the BCR-ABL tyrosine kinase in chronic myeloid leukemia. *N Engl J Med* **344**: 1031–1037.

Dubrez, L., Eymin, B., Sordet, O., Droin, N., Turhan, A.G., and Solary, E. (1998). BCR-ABL delays apoptosis upstream of procaspase-3 activation. *Blood* **91**: 2415–2422.

Duggan, D.J., Bittner, M., Chen, Y., Meltzer, P., and Trent, J.M. (1999). Expression profiling using cDNA microarrays. *Nat Genet* **21**: 10–14.

Enright, A.J., Iliopoulos, I., Kyrpides, N.C., and Ouzounis, C.A. (1999). Protein interaction maps for complete genomes based on gene fusion events. *Nature* **402**: 86–90.

Frueh, F.W. (2006). Impact of microarray data quality on genomic data submissions to the FDA. *Nat Biotechnol* **24**: 1105–1107.

Goetz, A.W., van der Kuip, H., Maya, R., Oren, M., and Aulitzky, W.E. (2001). Requirement for Mdm2 in the survival effects of Bcr-Abl and interleukin 3 in hematopoietic cells. *Cancer Res* **61**: 7635–7641.

Golub, T.R., Slonim, D.K., Tamayo, P., Huard, C., Gaasenbeek, M., Mesirov, J.P., Coller, H., Loh, M.L., Downing, J.R., Caligiuri, M.A., *et al.* (1999). Molecular classification of cancer: class discovery and class prediction by gene expression monitoring. *Science* **286**: 531–537.

Gorre, M.E., Mohammed, M., Ellwood, K., Hsu, N., Paquette, R., Rao, P.N., and Sawyers, C.L. (2001). Clinical resistance to STI-571 cancer therapy caused by BCR-ABL gene mutation or amplification. *Science* **293**: 876–880.

Guo, L., Lobenhofer, E.K., Wang, C., Shippy, R., Harris, S.C., Zhang, L., Mei, N., Chen, T., Herman, D., Goodsaid, F.M., *et al.* (2006). Rat toxicogenomic study reveals analytical consistency across microarray platforms. *Nat Biotechnol* **24**: 1162–1169.

Hacia, J.G. (1999). Resequencing and mutational analysis using oligonucleotide microarrays. *Nat Genet* **21**: 42–47.

Han, J., and Kamber, M. (2001). *Data Mining: Concepts and Techniques.* Academic Press, San Diego.

Hand, D., Mannila, H., and Smyth, P. (2001). *Principles of Data Mining.* MIT Press, London.

Hariharan, I.K., Adams, J.M., and Cory, S. (1988). bcr-abl oncogene renders myeloid cell line factor independent: potential autocrine mechanism in chronic myeloid leukemia. *Oncogene Res* **3**: 387–399.

Hedayat, A.S., Sloan, N.J.A., and Stufken, J. (1999). *Orthogonal Arrays — Theory and Applications.* Springer, New York.

Honda, H., and Hirai, H. (2001). Model mice for BCR/ABL-positive leukemias. *Blood Cells Mol Dis* **27**: 265–278.

Horst, R., and Pardalos, P.M. (1995). *Handbook of Global Optimization.* Kluwer Academic Publishers, Dordrecht, The Netherlands.

Hu, Y., Liu, Y., Pelletier, S., Buchdunger, E., Warmuth, M., Fabbro, D., Hallek, M., Van Etten, R.A., and Li, S. (2004). Requirement of Src kinases Lyn, Hck and Fgr for BCR-ABL1-induced B-lymphoblastic leukemia but not chronic myeloid leukemia. *Nat Genet* **36**: 453–461.

Hu, Y., Swerdlow, S., Duffy, T.M., Weinmann, R., Lee, F.Y., and Li, S. (2006). Targeting multiple kinase pathways in leukemic progenitors and stem cells is essential for improved treatment of Ph⁺ leukemia in mice. *Proc Natl Acad Sci USA* **103**: 16870–16875.

Ji, H., and Davis, R.W. (2006). Data quality in genomics and microarrays. *Nat Biotechnol* **24**: 1112–1113.

Jonuleit, T., van der Kuip, H., Miething, C., Michels, H., Hallek, M., Duyster, J., and Aulitzky, W.E. (2000). Bcr-Abl kinase down-regulates cyclin-dependent kinase inhibitor p27 in human and murine cell lines. *Blood* **96**: 1933–1939.

Kantardzic, M. (2002). *Data Mining: Models, Methods, and Algorithms*. Wiley Interscience–IEEE Press, Hoboken, NJ.

Khan, J., Wei, J.S., Ringner, M., Saal, L.H., Ladanyi, M., Westermann, F., Berthold, F., Schwab, M., Antonescu, C.R., Peterson, C., *et al.* (2001). Classification and diagnostic prediction of cancers using gene expression profiling and artificial neural networks. *Nat Med* **7**: 673–679.

Lander, E.S., Linton, L.M., Birren, B., Nusbaum, C., Zody, M.C., Baldwin, J., Devon, K., Dewar, K., Doyle, M., FitzHugh, W., *et al.* (2001). Initial sequencing and analysis of the human genome. *Nature* **409**: 860–921.

le Coutre, P., Tassi, E., Varella-Garcia, M., Barni, R., Mologni, L., Cabrita, G., Marchesi, E., Supino, R., and Gambacorti-Passerini, C. (2000). Induction of resistance to the Abelson inhibitor STI571 in human leukemic cells through gene amplification. *Blood* **95**: 1758–1766.

Li, D. (2004). Global optimisation for optical coating design. In: *Proceedings of the 2004 Conferences in Internet Technologies and Applications*, Purdue, IN, July 8–11, pp. 8–11.

Li, D. (2006). http://www.scis.ecu.edu.au/dli/.

Li, D., and Nathan, B. (1996). Global optimization advances multivariable thin-film design. *Laser Focus World* **5**: 135–136.

Li, L., Darden, T.A., Weinberg, C.R., Levine, A.J., and Pedersen, L.G. (2001a). Gene assessment and sample classification for gene expression data using a genetic algorithm/*k*-nearest neighbor method. *Comb Chem High Throughput Screen* **4**: 727–739.

Li, L., Weinberg, C.R., Darden, T.A., and Pedersen, L.G. (2001b). Gene selection for sample classification based on gene expression data: study of sensitivity to choice of parameters of the GA/KNN method. *Bioinformatics* **17**: 1131–1142.

Li, S., Ilaria, R.L., Jr., Million, R.P., Daley, G.Q., and Van Etten, R.A. (1999). The P190, P210, and p230 forms of the *BCR/ABL* oncogene induce a similar chronic myeloid leukemia-like syndrome in mice but have different lymphoid leukemogenic activity. *J Exp Med* **189**: 1399–1412.

Lionberger, J.M., Wilson, M.B., and Smithgall, T.E. (2000). Transformation of myeloid leukemia cells to cytokine independence by Bcr-Abl is suppressed by kinase-defective Hck. *J Biol Chem* **275**: 18581–18585.

Lipshutz, R.J., Fodor, S.P., Gingeras, T.R., and Lockhart, D.J. (1999). High density synthetic oligonucleotide arrays. *Nat Genet* **21**: 20–24.

Liu, J.J., Cutler, G., Li. W., Pan, Z., Peng, S., Hoey, T., Chen, L., and Ling, X.B. (2005). Multiclass cancer classification and biomarker discovery using GA-based algorithms. *Bioinformatics* **21**(11): 2691–2697.

Liu, R., Wang, X., Chen, G.Y., Dalerba, P., Gurney, A., Hoey, T., Sherlock, G., Lewicki, J., Shedden, K., and Clarke, M.F. (2007). The prognostic role of a gene signature from tumorigenic breast-cancer cells. *N Engl J Med* **356**: 217–226.

Loh, W. (1996). A combinatorial central limit theorem for randomized orthogonal array sampling designs. *Ann Stat* **24**: 1209–1224.

Mahon, F.X., Deininger, M.W., Schultheis, B., Chabrol, J., Reiffers, J., Goldman, J.M., and Melo, J.V. (2000). Selection and characterization of *BCR-ABL* positive cell lines with differential sensitivity to the tyrosine kinase inhibitor STI571: diverse mechanisms of resistance. *Blood* **96**: 1070–1079.

Majewski, M., Nieborowska-Skorska, M., Salomoni, P., Slupianek, A., Reiss, K., Trotta, R., Calabretta, B., and Skorski, T. (1999). Activation of mitochondrial Raf-1 is involved in the antiapoptotic effects of Akt. *Cancer Res* **59**: 2815–2819.

Marcotte, E.M., Pellegrini, M., Thompson, M.J., Yeates, T.O., and Eisenberg, D. (1999). A combined algorithm for genome-wide prediction of protein function. *Nature* **402**: 83–86.

Marley, S.B., Deininger, M.W., Davidson, R.J., Goldman, J.M., and Gordon, M.Y. (2000). The tyrosine kinase inhibitor STI571, like interferon-alpha, preferentially reduces the capacity for amplification of granulocyte-macrophage progenitors from patients with chronic myeloid leukemia. *Exp Hematol* **28**: 551–557.

McGahon, A.J., Nishioka, W.K., Martin, S.J., Mahboubi, A., Cotter, T.G., and Green, D.R. (1995). Regulation of the Fas apoptotic cell death pathway by Abl. *J Biol Chem* **270**: 22625–22631.

Mitra, S., and Acharya, T. (2005). *Data Mining: Multimedia, Soft Computing, and Bioinformatics*. John Wiley & Sons, Newark, NJ.

Mitra, S., Pal, S.K., and Mitra, P. (2002). Data mining in soft computing frame-work: a survey. *IEEE Trans Neural Netw* **13**: 3–14.

Montgomery, D.C. (1997). *Design and Analysis of Experiments*, 4th ed. Wiley, New York.

Neshat, M.S., Raitano, A.B., Wang, H.G., Reed, J.C., and Sawyers, C.L. (2000). The survival function of the Bcr-Abl oncogene is mediated by Bad-dependent and -independent pathways: roles for phosphatidylinositol 3-kinase and Raf. *Mol Cell Biol* **20**: 1179–1186.

O'Hare, T., Pollock, R., Stoffregen, E.P., Keats, J.A., Abdullah, O.M., Moseson, E.M., Rivera, V.M., Tang, H., Metcalf, C.A., 3rd, Bohacek, R.S., *et al.* (2004). Inhibition of wild-type and mutant Bcr-Abl by AP23464, a potent ATP-based oncogenic protein kinase inhibitor: implications for CML. *Blood* **104**: 2532–2539.

Ooi, C.H., and Tan, P. (2003). Genetic algorithms applied to multi-class prediction for the analysis of gene expression data. *Bioinformatics* **19**: 37–44.

Owen, A.B. (1992). Orthogonal arrays for computer experiments: integration and visualization. *Stat Sin* **2**: 439–452.

Owen, A.B. (1994). Lattice sampling revisited: Monte Carlo variance of means over randomized orthogonal arrays. *Ann Stat* **22**: 930–945.

Pane, F., Frigeri, F., Sindona, M., Luciano, L., Ferrara, F., Cimino, R., Meloni, G., Saglio, G., Salvatore, F., and Rotoli, B. (1996). Neutrophilic-chronic myeloid leukemia: a distinct disease with a specific molecular marker (*BCR/ABL* with C3/A2 junction). *Blood* **88**: 2410–2414.

Parada, Y., Banerji, L., Glassford, J., Lea, N.C., Collado, M., Rivas, C., Lewis, J.L., Gordon, M.Y., Thomas, N.S., and Lam, E.W. (2001). BCR-ABL and inter-leukin 3 promote haematopoietic cell proliferation and survival through mod-ulation of cyclin D2 and p27Kip1 expression. *J Biol Chem* **276**: 23572–23580.

Patterson, T.A., Lobenhofer, E.K., Fulmer-Smentek, S.B., Collins, P.J., Chu, T.M., Bao, W., Fang, H., Kawasaki, E.S., Hager, J., Tikhonova, I.R., *et al.* (2006). Performance comparison of one-color and two-color platforms within the MicroArray Quality Control (MAQC) project. *Nat Biotechnol* **24**: 1140–1150.

Peng, S., Xu, Q., Ling, X.B., Peng, X., Du, W., and Chen, L. (2003). Molecular classification of cancer types from microarray data using the combination of genetic algorithms amd support vector machines. *FEBS Lett* **555**: 358–362.

Perou, C.M., Jeffrey, S.S., van de Rijn, M., Rees, C.A., Eisen, M.B., Ross, D.T., Pergamenschikov, A., Williams, C.F., Zhu, S.X., Lee, J.C., *et al.* (1999). Distinctive gene expression patterns in human mammary epithelial cells and breast cancers. *Proc Natl Acad Sci USA* **96**: 9212–9217.

Rao, C.R. (1946). Hypercubes of strength *d* leading to confounded designs in factorial experiments. *Bull Calcutta Math Soc* **38**: 67–78.

Rao, C.R. (1947). Factorial experiments derivable from combinatorial arrangements of arrays. *J R Stat Soc* **9**(Suppl): 128–139.

Rao, C.R. (1949). On a class of arrangements. *Proc Edinb Math Soc* **8**: 119–125.

Sanchez-Garcia, I., and Martin-Zanca, D. (1997). Regulation of Bcl-2 gene expression by BCR-ABL is mediated by Ras. *J Mol Biol* **267**: 225–228.

Sawyers, C.L. (1997). Signal transduction pathways involved in BCR-ABL transformation. *Baillieres Clin Haematol* **10**: 223–231.

Sawyers, C.L. (1999). Chronic myeloid leukemia. *N Engl J Med* **340**: 1330–1340.

Schena, M., Shalon, D., Davis, R.W., and Brown, P.O. (1995). Quantitative monitoring of gene expression patterns with a complementary DNA microarray. *Science* **270**: 467–470.

Shah, N.P., Nicoll, J.M., Nagar, B., Gorre, M.E., Paquette, R.L., Kuriyan, J., and Sawyers, C.L. (2002). Multiple *BCR-ABL* kinase domain mutants confer polyclonal resistance to the tyrosine kinase inhibitor imatinib (STI571) in chronic phase and blast crisis chronic myeloid leukemia. *Cancer Cell* **2**: 117–125.

Shah, N.P., Tran, C., Lee, F.Y., Chen, P., Norris, D., and Sawyers, C.L. (2004). Overriding imatinib resistance with a novel ABL kinase inhibitor. *Science* **305**: 399–401.

Shi, L., Reid, L.H., Jones, W.D., Shippy, R., Warrington, J.A., Baker, S.C., Collins, P.J., de Longueville, F., Kawasaki, E.S., Lee, K.Y., *et al.* (2006). The MicroArray Quality Control (MAQC) project shows inter- and intraplatform reproducibility of gene expression measurements. *Nat Biotechnol* **24**: 1151–1161.

Shi, Q., Abbruzzese, J.L., Huang, S., Fidler, I.J., Xiong, Q., and Xie, K. (1999). Constitutive and inducible interleukin 8 expression by hypoxia and acidosis renders human pancreatic cancer cells more tumorigenic and metastatic. *Clin Cancer Res* **5**: 3711–3721.

Shippy, R., Fulmer-Smentek, S., Jensen, R.V., Jones, W.D., Wolber, P.K., Johnson, C.D., Pine, P.S., Boysen, C., Guo, X., Chudin, E., *et al.* (2006). Using RNA sample titrations to assess microarray platform performance and normalization techniques. *Nat Biotechnol* **24**: 1123–1131.

Skorski, T., Bellacosa, A., Nieborowska-Skorska, M., Majewski, M., Martinez, R., Choi, J.K., Trotta, R., Wlodarski, P., Perrotti, D., Chan, T.O., *et al.* (1997). Transformation of hematopoietic cells by BCR/ABL requires activation of a PI-3k/Akt-dependent pathway. *EMBO J* **16**: 6151–6161.

Skorski, T., Nieborowska-Skorska, M., Wlodarski, P., Perrotti, D., Martinez, R., Wasik, M.A., and Calabretta, B. (1996). Blastic transformation of p53-deficient bone marrow cells by p210bcr/abl tyrosine kinase. *Proc Natl Acad Sci USA* **93**: 13137–13142.

Steffen, J., von Nickisch-Rosenegk, M., and Bier, F.F. (2005). *In vitro* transcription of a whole gene on a surface-coupled template. *Lab Chip* **5**: 665–668.

Storey, J.D. (2002). A direct approach to false discovery rates. *J R Stat Soc Ser B Stat Methodol* **64**: 479–498.

Talpaz, M., Sawyers, C.L., Kantarjain, H., Resta, D., Fernandes Rees, S., Ford, J., and Bruker, B.J. (2000). Activity of an ABL specific tyrosine kinase inhibitor in patients with BCR/ABL positive acute leukemias, including chronic myelogenous leukemia in blast crisis. *Oncologist* **5**: 282–283.

Tong, W., Lucas, A.B., Shippy, R., Fan, X., Fang, H., Hong, H., Orr, M.S., Chu, T.M., Guo, X., Collins, P.J., *et al.* (2006). Evaluation of external RNA controls for the assessment of microarray performance. *Nat Biotechnol* **24**: 1132–1139.

Tou, J.T., and Gonzalez, R.C. (1974). *Pattern Recognition Principles.* Addison-Wesley, London.

Venter, J.C., Adams, M.D., Myers, E.W., Li, P.W., Mural, R.J., Sutton, G.G., Smith, H.O., Yandell, M., Evans, C.A., Holt, R.A., *et al.* (2001). The sequence of the human genome. *Science* **291**: 1304–1351.

von Bubnoff, N., Schneller, F., Peschel, C., and Duyster, J. (2002). BCR-ABL gene mutations in relation to clinical resistance of Philadelphia-chromosome-positive leukaemia to STI571: a prospective study. *Lancet* **359**: 487–491.

Wang, L., and Fu, X. (2005). *Data Mining with Computational Intelligence.* Springer-Verlag, Berlin.

Warmuth, M., Bergmann, M., Priess, A., Hauslmann, K., Emmerich, B., and Hallek, M. (1997). The Src family kinase Hck interacts with Bcr-Abl by a kinase-independent mechanism and phosphorylates the Grb2-binding site of Bcr. *J Biol Chem* **272**: 33260–33270.

Weisberg, E., and Griffin, J.D. (2000). Mechanism of resistance to the ABL tyrosine kinase inhibitor STI571 in BCR/ABL-transformed hematopoietic cell lines. *Blood* **95**: 3498–3505.

Woelfle, U., Cloos, J., Sauter, G., Riethdorf, L., Janicke, F., van Diest, P., Brakenhoff, R., and Pantel, K. (2003). Molecular signature associated with bone marrow micrometastasis in human breast cancer. *Cancer Res* **63**: 5679–5684.

Wren, J.D., Yao, M., Langer, M., and Conway, T. (2004). Simulated annealing of microarray data reduces noise and enables cross-experimental comparisons. *DNA Cell Biol* **23**: 695–700.

Wu, H., Kerr, K., and Churchill, G.A. (2003). MAANOVA: a software package for the analysis of spotted cDNA microarray experiments. In: Parmigiani, G., Garrett, E., Irizarry, R., and Zeger, S. (eds.), *The Analysis of Gene Expression Data: Methods and Software*, Springer-Verlag, New York, pp. 313–341.

Zhang, S.D., and Gant, T.W. (2004). A statistical framework for the design of microarray experiments and effective detection of differential gene expression. *Bioinformatics* **20**: 2821–2828.

Index